HOUSING POLICY AND FINANCE

HOUSING

POLICY

AND

FINANCE

JOHN BLACK AND DAVID C. STAFFORD

ROUTLEDGE
London & New York

First published in 1988 by
Routledge
11 New Fetter Lane, London EC4P 4EE

Published in the USA by
Routledge
in association with Routledge, Chapman & Hall, Inc.
29 West 35th Street, New York, NY 10001

© 1988 J. Black and D.C. Stafford

Photoset by Mayhew Typesetting, Bristol, England
Printed and bound in Great Britain
by Billing & Sons Limited, Worcester.

British Library Cataloguing in Publication Data

Black, John, 1931–
    Housing policy and finance.
    1. Housing policy — Great Britain
    I. Title   II. Stafford, David C.
    338.4′7363556′0941   HD7333.A3

    ISBN 0–415–00419–5
    ISBN 0–415–00420–9 Pbk.

Library of Congress Cataloging-in-Publication Data

    ISBN 0–415–00419–5
    ISBN 0–415–00420–9 Pbk.

Contents

Figures   vii

Tables   viii

Preface   ix

Part I   Housing Facts and Fiction

1. The Political Economy of Housing   3
   General aims of housing policy   3
   Characteristics of housing   6
   Supply and demand for rented housing   11
   Alternative systems   12

2. The Facts about Housing in the UK   17
   Investment in housing   17
   Changes in number of dwellings and households   20
   Changes in condition of the housing stock   21
   Changes in tenure   26
   Finance of housing — private and public sector   28

Part II   Housing Tenure and Finance

3. Owner-occupation: Choice and Finance   37
   Preferences for owner-occupation   37
   Finance for owner-occupation   39
   Drawbacks with owner-occupation   44

4. Public-sector Housing: Finance and Allocation   46
   Aims of public-sector housing   46
   Local authority housing finance   47
   Rents and subsidies   50
   Allocation of local authority housing   56
   Drawbacks with council housing   60

5. Private Rented Housing: Rent Control in Theory and
   Practice   63
   Decline and fall   63
   Theoretical models of rent control   68
   Consequences of rent control   74

Part III Housing Finance and Policy Reforms

6. Taxation and Owner-occupied Housing   87
     Mortgage interest tax relief   87
     Taxation of imputed income   94
     Capital gains tax exemption   98

7. Reform of Council Housing   101
     The issues   101
     Methods of privatisation   103
     Limitations to council house sales   106
     Future of council housing   111
     Housing Associations   112
     Future policy   114

8. Policy Options for Private Rented Housing   116
     Municipalisation   116
     Policy options under conditions of rent control   117
     Expansion of the voluntary sector   122
     Abolition of rent and eviction controls   123

9. Conclusions   129
     Council housing   129
     Owner-occupied housing   131
     Private rented housing   131
     Tenure neutrality   132

Bibliography   135

Index   140

Figures

2.1 Households and dwellings, 1951–86   21

2.2 Stock of dwellings in the UK, 1951–85 (by tenure)   27

2.3 Public expenditure on housing in the UK, 1980–7   30

5.1 The market for housing   69

5.2 (a), (b) Equilibrium in the housing market   73

8.1 What is a fair rent?   119

8.2 Fair rents with increasing costs   119

8.3 The effects of housing subsidies   121

8.4 Redistribution from rent control   126

Tables

2.1 Investment in dwellings, 1975–85, at 1980 prices   18

2.2 Private sector investment in dwellings, 1975–85, at 1980 prices   18

2.3 Public sector investment in dwellings, 1975–85, at 1980 prices   19

2.4 Housing conditions in Great Britain, 1971–84 (availability of amenities)   22

2.5 Housing standards in Great Britain, 1971 and 1984 (by tenure)   24

2.6 Distribution of dwellings in the UK, 1914–85 (by tenure)   26

3.1 Net mortgage advances in the UK, 1971–84 (by source)   41

4.1 Summary of local authority Housing Revenue Accounts, 1984–5 (estimated)   49

4.2 Council house subsidies, 1975–85   50

4.3 Local and central government support for housing in the UK, 1979–80 and 1983–4   53

4.4 Gross unrebated rents, 1978–86 (average amount in per dwelling per week)   54

4.5 Allocation of local authority housing in England and Wales, 1971–85   57

6.1 Growth of mortgage interest tax relief, 1979–86   89

6.2 Mortgage interest tax relief, 1983–4 (by income bands)   89

6.3 'Leakage' from the private housing market, 1979–84 (estimates of net cash withdrawals)   91

6.4 Personal income tax deductibility in selected countries   92

7.1 Sales of dwellings owned by local authorities and new towns, 1971–84, England and Wales; and 1980–4, United Kingdom   104

Preface

This book seeks to examine the economics of housing policy and finance. The importance of housing can hardly be exaggerated. In human terms, access to satisfactory housing is vital to health, happiness and civilised living. The economic facts about housing bear this out. In the UK in 1985, ownership of dwellings gave rise to 5.8 per cent of the gross domestic product, and use of dwellings accounted for 9.2 per cent of total consumer expenditure. In both cases the figure excludes payment of rates and spending on home improvements, but includes 'imputed incomes' from the use of owner-occupied houses.

In 1985 dwellings accounted for 20.2 per cent of gross fixed capital formation, *i.e.* of investment, in the UK, and for 31.6 per cent of net investment; the net figure is high because housing wears out more slowly than most capital assets. Over the years 1975-85, dwellings accounted for 20.4 per cent of total fixed investment, so that 1985 was quite normal in this respect. At the end of 1985 the value of the stock of dwellings, at replacement cost, was 30.7 per cent of the total real domestic capital of the UK.

Thus housing accounts for a very significant part of total economic activity in the UK, and is of great importance for welfare. It is therefore extremely important that this large quantity of resources should be used efficiently. This book sets out to examine whether the nation's housing stock is being used efficiently, and to consider how public policies towards housing have affected the way housing is used, and whether these policies could be improved.

In advocating a greater use of the market in housing policy, we are motivated by a genuine belief that this would improve housing standards.

This book has been written with much encouragement from friends and colleagues. In particular, we would like to thank Dr Robert Albon of the Australian National University, Canberra, for his support and advice, and Diccon Pearse of Exeter City Council Treasurer's Department for his valuable and extensive comments. Marjorie Anne Howe and Sylvia Northam at the University of Exeter deserve especial praise and thanks for typing the draft. Of course, we are responsible for the contents and opinions.

Part I

Housing Facts and Fiction

1

The Political Economy of Housing

GENERAL AIMS OF HOUSING POLICY

One could summarise the general aim of housing policy by saying that one would like to see everybody housed at a standard consistent with civilised living, and in an acceptable location. What constitutes an acceptable standard will depend on the standard generally attained in a country. For the UK this would include sufficient rooms for decency, adequate cooking and washing arrangements, and a normal state of repair. The point about location is that people have needs which can only be satisfied by living in acceptable places. For some people this means the ability to continue living in an area where they have family and friends within easy reach. For others, it means the ability to move in pursuit of employment. If areas change in relative popularity this may give rise to problems — it may not be possible for everybody who would like to live in a particular area to do so.

Organisation of housing

There are several possible methods by which housing could be organised and allocated. These have varied merits and drawbacks, and it is probable that no single method on its own could allow the aims of housing policy to be achieved.

One method is owner-occupation. This has the great attraction that an owner has more discretion over the quality of a dwelling, through control over maintenance and improvements, than a tenant. Ownership usually has the advantage that one can choose to stay put if one wishes (subject only to the possibility of compulsory public

3

purchase for some purpose such as road building). Also one can move if one wants to, since a property can be sold and the proceeds spent on buying a house elsewhere.

The main disadvantage of ownership is that the owner is left with the entire financial responsibility for maintenance and repairs, and the burden of organising them. This can be very expensive, and many items of repairs are not insurable. An owner-occupier also has to have a lot of capital tied up in a house, which may be impossible if one's total assets are small, and awkward even if they are just large enough to make it possible.

A second method is that ownership and occupation should be separated, houses being owned by one body, either private or public, which is responsible for maintenance and improvements, and for providing the necessary capital; and occupied by another who then pays rent. This system is convenient for many types of household: for those who are too poor to be able to afford to own a house; for those who are unsuited through age, disability, or mere lack of energy and intelligence, to take responsibility for repairs; and for those whose careers involve frequent mobility, who would find the costs and trouble of frequent sales and purchases burdensome.

The disadvantage of renting is lack of control over the quality of the accommodation, and a lack of security of tenure if the landlord decides that the property could more profitably be used for some other purpose, or merely wants to let it to somebody else. A third disadvantage, from the point of view of would be tenants, is that landlords will not want to let to people whom they expect, for whatever reason, to be likely to fail to pay rents promptly or to misuse the property. The grounds for such expectations may vary from the purely factual, *i.e.* past bad experience with the same people, to mere prejudice against 'alien' groups. People who are regarded, for whatever reason, as likely to be arsonists, vandals, or dilatory in paying the rent, are unlikely to find willing private landlords. People who are expected to make unsatisfactory tenants will probably need to be housed by public authorities if they are to be decently housed at all.

Which form of housing tenure people choose, both as occupiers and property owners, will be affected in various ways by public policies.

Public policy affects owner-occupation through various channels. The tax system affects both people's ability and desire to be owner-occupiers. Owning dwellings gives rise to a stream of incomes — if they are let out, this will be rent and the landlord will pay tax on it. If owners occupy their houses themselves there is a similar stream of real services, which accrue to the owner-occupier as an 'imputed' income. The tax system could treat these incomes like any other and tax them; or it could tax them on some special basis; or it could ignore them, as is currently done in the UK. The more leniently the imputed income of owner-occupiers is taxed, the more attractive it will be to become an owner-occupier. Similar decisions have to be made as to how to treat owner-occupied housing in any taxes which are levied on wealth, capital gains or capital transfers.

Many people become owner-occupiers without themselves having sufficient wealth to purchase a house outright. This is done by borrowing the money on a mortgage; the house is pledged as collateral for a loan, and the borrower pays interest and pays off the principal gradually, often over 25 years. The attractiveness of owner-occupation will be affected by the way in which the tax system treats these interest and redemption payments. In the UK interest on the first £30,000 of a mortgage is tax deductible, but redemption payments, and interest on the part of a mortgage in excess of £30,000 are not deductible.

A second way in which public policy can affect housing is by the provision of housing by the public sector. Public bodies, such as central or local government, or special institutions such as new town corporations in the UK, can buy or build housing for the purposes of letting it. They then have to decide how much rent to charge and what conditions of tenancy to offer, and how to select tenants. Clearly the lower the rents and the more generous the conditions of tenure, the more attractive renting houses from the public sector will be. If the rents are sufficiently low and the conditions sufficiently attractive there will be excess demand for public authority housing. If the authorities decide that they should let to those in the greatest need, they need to decide how need should be assessed. They must also weigh up the balance of advantage between security of tenure for existing tenants and availability of houses to meet the needs of those applying for tenancies. Lower rents, better conditions and more secure tenure for sitting tenants mean fewer vacancies for anybody else.

A third aspect of public policies is control of the rents and conditions of private letting, and the tax treatment of private landlords and tenants.

It would be possible, at one extreme, for the government to leave decisions about rents, conditions and tenure entirely to the market. Landlords would then offer tenants whatever combination of rent,

5

conditions and tenure provisions seemed most profitable; tenants who wanted better maintenance and more secure tenures would have to pay more for them.

Alternatively, the government could seek to regulate the terms of private tenancies; this could comprise restrictions on the level of rents, provisions about minimum standards of maintenance, and regulations concerning security of tenure. Such controls would have effects on the supply of and demand for rented housing, which need to be investigated in detail to predict the effects of controls.

CHARACTERISTICS OF HOUSING

An analysis of the housing market requires a broad elaboration of its structure and mechanism and the influences on demand and supply in the determination of the production, allocation and price of dwellings. While income, tastes and household and financial circumstances determine the demand for the quality and quantity of housing, the resultant stock and standard acquired represents an important contributor to the growth, development and welfare of a society. For these reasons, it is not surprising that housing has always been subject to considerable political controversy; it ranks with education and the health services as a focal point of social provision.

Durability and cost

Housing is a highly durable asset and, once produced, is wholly immobile. Moreover, new building in any one year represents such a small percentage of total stock that it is difficult for supply to respond in the short run to changes in demand except through house conversion or improvements.

While housing is generally defined for statistical purposes as dwelling units (housing units occupied separately by households), it comprises a great variety of qualities and quantities. Indeed, the spectrum of housing units embraces shacks and ducal palaces, as well as the more familiar classifications of detached houses, bungalows, semi-detached houses, flats and maisonettes, *etc.*. Moreover, these classifications hardly do justice to the very wide variety of types of housing units that exist.

The high capital cost of producing a dwelling unit is such that it

6

represents a multiple of yearly income for most purchasers, who must negotiate a mortgage for its acquisition, with capital repayments and interest costs until maturity of the loan. An individual must normally borrow finance and the relevant requirements of given income, health and collateral are such that borrowing capacities vary widely between potential borrowers. Income tax allowances on mortgage interest and capital gains tax exemption can also affect housing demand and supply, together with the considerable influence of financial institutions via their lending conditions.

Stocks and flows

As a durable good, housing provides both necessary and luxury flows of consumption services, as well as constituting a stock or capital asset to its owner. We may distinguish between two basic housing markets.

First, there is the market for housing services. This is a consumer's market, whereby each dwelling unit is presumed to yield some quantity and quality of services in each time period. These services have diverse quantitative and qualitative aspects, and a consumer, or household, can trade one service against another to derive the best package of services at a given price or rent level.

Second, there is the market for housing stocks. This is the investment market. While the owner-occupier is in both markets as an asset-holder and also as a beneficiary of housing services, a landlord is in this market as an asset-holder, and a tenant in the former market as a consumer of housing services.

Although these markets can be usefully distinguished, the factors which determine the price of any capital asset and the rent of its services are likely to be similar. If, for example, the stock of housing is changed through new building or clearance, then there will also be a change in the consumption of housing services.

The respective demands for housing services and stocks are clearly related but should be distinguished.

The demand for housing services depends on the price of alternative accommodation, the prices of other goods, the number of households, the household income, tastes and ages, household size and composition, and the availability and terms of finance. Other variables, of less direct but possibly of similar importance, will include its location relative to a town or neighbourhood centre, proximity of amenities and jobs, and demographic, physical and social environmental factors.

The demand for housing stock is clearly derived from the demand for housing services, for if the stock did not provide any service it would have no value in the housing market. To the landlord, the services of the investment are reflected in a monetary yield in the rents received, while the owner-occupier derives a non-monetary yield in the space, quality and other characteristics the house provides in services.

Housing stock will therefore be purchased provided the return on the capital sum invested is greater than the return on alternative investment opportunities. The relative performance of alternative investments will influence the holding of portfolio assets and if yields on, say, securities, provide greater overall returns than housing, owners of houses may be tempted to adjust the quantity of their stock of housing assets and switch to more profitable outlets.

While landlords of residential property are more sensitive to investment returns and may attempt (or would wish in the absence of present Rent Acts) to sell their property or reduce the quality of housing service to compensate for a fall in investment return, owner-occupiers may be tempted to adjust their holding of housing stock downwards.

By contrast, when there are expectations of rising values then not only may new entrants and landlords be attracted to house ownership, but existing owner-occupiers may be tempted to increase their stock of housing, either by purchasing a more expensive property or by increasing the size of their house.

In either case, therefore, housing services and stocks are related as the increased stock will provide additional housing services. In short, in the housing market, the price of housing services per year should reflect the costs of using a given housing unit in that year, *i.e.* maintenance, repairs, taxes and the opportunity cost of capital.

Environmental factors

Location and environmental factors are important in housing. The benefits people get from a dwelling, and hence the amount they are willing to pay to buy or rent it, depend partly on its own characteristics, such as plot size, number and size of rooms, standard of equipment, quality of design, *etc.* The value of a dwelling is also influenced by its location, *i.e.* how close it is to workplaces, shops, schools, public transport, *etc.* It is also influenced by environmental factors. Values will be high, for given accommodation, if the neighbourhood is law-abiding, clean, quiet and unpolluted, and the residents devoted to gardening. Values will be low, for given accommodation, if the neighbourhood is crime-ridden, noisy, dirty and subject to vandalism.

The environment will over the course of time affect the quality of homes available in a neighbourhood. The maintenance and improvement of dwellings cost money, and investment will offer a better return if a neighbourhood is pleasant and expected to improve than if it is nasty and expected to get worse. Thus areas which are already pleasant attract more expenditure on maintenance and improvements. Areas which are below average may enter a vicious circle, in which their unattractiveness deters people from spending money on improvements, and the lack of maintenance combines with vandalism to make conditions deteriorate so that they end up as slums.

In the face of the emergence of run-down, poor quality housing, much post-war policy for the housing of the poor has been heavily concentrated on purpose-built council housing of a certain minimum standard to provide cheap, standardised housing units. This policy, as discussed later, has not been without defects. For example, the purpose-built accommodation has often been of only just adequate standard, so that a combination of hard usage and gradually rising standards means that it soon comes to be regarded as slums.

An alternative approach to improvement is through, for example, the use of improvement grants for houses which have 'filtered down'. In studies of market operations, a good deal of analysis has been directed towards the concept of 'filtering', which refers to the process of changes in occupancy as a property gradually becomes available to those of lower income.

Housing shortage

What is meant by saying there is a shortage of housing? One approach is to compare the number of households with the number of dwellings; if there are more households than dwellings then some people must be homeless, and there is obviously a shortage of housing. This however is a far too simple approach. The stock of housing varies greatly in location, size, condition and price. In the UK some people are homeless even though the total stock of dwellings outnumbers households by over a million. This is because, while the overall stock of dwellings is more than adequate, there are not enough in some areas at prices people can afford. There is also the problem that many people who do have somewhere to live are occupying dwellings which they or other people regard as being overcrowded, in an unsatisfactory condition or inconveniently located.

If the price of dwellings were free to fluctuate so as to clear the market, there could still be shortages in the sense that dwellings were overcrowded or in unsatisfactory condition, or too far from people's work or families. If rents are controlled, there can also be shortages in the sense of excess demand; more people would like to rent homes at the controlled price than there are homes available.

If there is a shortage of housing in any given area, what can be done about it? There are several possible approaches, and we will later consider the potential for these, and their limitations.

(a) The stock of dwellings can be increased, by building or conversion, or change of type of tenure.

(b) The demand for homes in an area can be reduced, by the use of non-housing policies such as control on the building of factories, shops and offices and other sources of jobs in overcrowded areas.

(c) If excess demand is due to controlled rents, increasing them may reduce local shortages by inducing people who can do so to live elsewhere. On the other hand higher rents will worsen the housing problems of those who can't conveniently move, and may force them to accept overcrowding or sub-standard accommodation.

(d) Where housing problems are due to low incomes, then either general income support or specific help with housing costs can enable people to afford decent accommodation, if it is available.

(e) This leaves the problem of people who are homeless or badly housed because of their own tastes or handicaps. There are people who even if they had, or were given, the means to afford adequate housing, would prefer to spend the money on other things — drink, drugs and gambling spring to mind. There are also elderly people whose housing is unsatisfactory because they are unable to organise repairs. For these groups ending housing shortages may require legal controls, for example on overcrowding or the condition of dwellings, or public provision of housing.

SUPPLY AND DEMAND FOR RENTED HOUSING

Let us consider the possible sources of changes in the supply and demand for rented housing in any particular area. Most of the possible changes take time to organise, so that supply and demand are both likely to be more responsive to changes in price the longer the time horizon we consider. However, some variation is possible even in the fairly short run.

Supply

Take first the supply of houses; this can vary in the following ways:
— Houses can be built or demolished.
— Derelict houses can be restored, or usable ones can be allowed to deteriorate by failure to repair them.
— Houses can be converted to other uses, including shops, offices, workshops and use for storage; or premises used for other purposes can be converted for housing.
— Houses previously let can be sold for owner-occupation, or houses previously owner-occupied can be rented out.
— Houses can be converted. Large houses can be divided into flats, or small houses can be combined to form better ones.
— People with large enough houses can sublet parts, or can cease to do so.
— People can take in lodgers, or cease to do so.

Demand

Demand can vary in the following ways:
— People can vary the amount of housing they occupy, moving into larger or smaller dwellings.
— People can move to other areas, or outsiders can move in.
— People can share with their parents, children, other relatives or friends, or can cease to do so.
— People can transfer to the owner-occupier market, or owner-occupiers can start to rent.

ALTERNATIVE SYSTEMS

There is a variety of possible systems for allocating people to dwellings and dwellings to people. We will analyse these to try to discover which of them is best at utilising the various possibilities of changing the supply and demand for housing, and what groups of the population gain and lose from each possible system. While actual systems may differ in minor ways from the extremes described below, the following three systems of allocation seem worthy of analysis.

Market allocation

In a pure market solution, each dwelling would be let at a rent, on conditions of maintenance and length of tenure decided by bargaining between landlords and tenants. For any given standard of maintenance and length of notice, each tenancy would go to the highest bidder.

From some points of view a market solution has merits. Landlords would like it. People with well paid jobs would be assured of mobility when they needed it in pursuit of even better jobs, as they could afford to overbid others for the houses of their choice. The same would apply to prosperous people who did not need to change area, but did want to change the character of their housing. For example, they might prefer a flat when young, a house with garden when they had children, and a move to a smaller house or flat with less garden when the children were grown up and they became too old to want to have to manage a larger home.

People who would do badly from a market system would be the poor, who could never afford nice houses, especially those without jobs, whether unemployed or retired. They would also find themselves liable to be priced out of whatever areas or types of house were currently in fashion with the better off.

A merit of the market system is that choices about moving area or changing size or type of housing would be taken by people for themselves; they have better information about their own needs and preferences than anybody else could have. Thus scarce types of housing would tend to be efficiently used.

12

Administrative allocation

An opposite extreme system would be one in which a public
authority — let us call it the local authority — had control of all
rented accommodation, and allocated people to dwellings and dwellings to people as it saw fit. This system would enable the housing
authority to ensure that some accommodation was allocated to
everybody it wanted to house. On the other hand people would only
get accommodation in an area if they could persuade the allocating
body to provide for them, and they would only get the kind of
accommodation they preferred if they could persuade the authority
to allocate it to them. That is, the system would favour any person
or group favoured by the authority, and penalise anybody not in
favour.

Also, if any household is allocated more or better accommodation
than it really wanted, there is no incentive to share it, and indeed it
may not even be allowed to.

While in principle a housing authority could favour any group at
the expense of any other, it is of interest to observe the actual
priorities used by local authorities in the UK. First, while there has
been no actual legal requirement of this, the rights of sitting tenants
have been widely respected. Families have been left in the same
housing even if their means and family size have changed drastically
since the tenancy was originally allocated to them. One can see the
motivation for this; any decision to turn them out would inevitably
be regarded as arbitrary by the victims and their friends, relations
and neighbours, and councillors prefer a quiet life. Second, in
allocating tenancies that do become vacant, councils have shown a
strong preference for people already living in the area, and have
usually preferred those who have been there a long time. It has been
very difficult for those outside an area to get tenancies in it. Again,
as those resident in the area have votes in it and those outside do not,
one can see the motivation.

If a system of allocation favours sitting tenants, and long-term
occupants of an area, it necessarily penalises those who are not
tenants already, or who would like to move into the area. These
groups will necessarily include the young who are trying to start new
households, and migrants. A low turnover rate of tenancies helps to
reduce the efficiency of use of housing when it is administratively
allocated.

13

A controlled market

The third system we will analyse is a controlled market. In this system the public authorities restrict rents and lay down minimum standards of maintenance and minimum periods of notice, possibly to the extent of permanent protection of tenure for tenants.

Clearly this system is bad for landlords. Existing tenants are favoured, especially when the system starts. On the other hand once rents are controlled the landlord has no incentive to maintain the property to any standard higher than the minimum laid down, which may in any case not be easy to enforce. Rent control may also mean that the landlord cannot afford maintenance. Thus conditions are likely to deteriorate to the minimum permitted by law, and possibly lower. This is more likely still if the landlord is able to sell with vacant possession to an owner-occupier if there is a vacancy, since the more uncomfortable the premises become the more likely the tenant is to move out. Alternatively tenants may have to take over some maintenance and make improvements themselves; however they only get the benefits so long as they stay, so if they are liable to have to move this course will not be attractive to them.

Both lack of maintenance, and the incentive for the landlord to sell when the chance occurs, make it likely that rent controls will lead to a gradual fall in the supply of rented accommodation. Protection of tenures also has the disadvantage that it will deter people from letting parts of their homes if they expect they will have trouble in getting rid of tenants who are unsatisfactory, or if they are liable to want to sell in order to move.

If a landlord does relet a vacant property subject to rent control, it will pay to be selective about who to choose as tenants. Landlords will prefer tenants who seem likely to pay the rent without trouble or delay, to avoid damage to the property, and to carry out maintenance and improvements for themselves. Landlords will try to avoid taking on as tenants people who are very poor, or who have large families, or who have bad habits such as drink, drugs or criminal records, as these are liable to lead to poor paying power and vandalism. Quite apart from economic motives to discriminate, if landlords have any prejudices about race, religion, language or sexual orientation, rent control allows them to discriminate at little cost to themselves.

Thus a controlled market system will be bad for newcomers to an area, or to the housing market as a whole, *i.e.* the young, and it will be especially bad for people who look likely to be unsatisfactory tenants. It will also penalise sitting tenants who would prefer to move to other areas in search of work, or who would like to change their type of accommodation due to changes in family circumstances, since they can only move at the cost of becoming outsiders.

Housing is unlikely to be efficiently allocated under a regulated market system; controlled rents mean that sitting tenants have no incentive to economise on housing; they can't move out of crowded areas or unsuitable accommodation because excess demand makes it difficult to get a tenancy anywhere else.

Mixed systems

It is probably unlikely that any one of these ideal systems will ever be used on its own. Actual systems will contain sections of the housing stock run on each system. The systems themselves will very probably not exist in their extreme form discussed above.

In an otherwise market-run sector, for example, public health considerations will probably cause some regulation of minimum standards of maintenance, while concern about the effects on tenants of the cost and trouble of frequent moves could be used to justify some degree of security of tenure, e.g. a minimum period of six months notice to quit.

An otherwise administrative allocation sector, too, would probably contain some market elements, the obvious example being collection of rents for council houses. Even if they are not fixed high enough to clear the market, rents will restrict the degree of excess demand for housing and so reduce the administrative problem of allocating vacancies, by raising vacancies above and reducing applications below the levels that would prevail with zero rents.

Conclusions

It seems fair to summarise the relative merits and demerits of the different allocation systems as follows. The market system has the merit of helping people who want to move, either geographically or between types of housing, to do so. It also has the merit of giving everybody an incentive to use housing efficiently. It is the best system for landlords, and thus promotes a continuing supply of rented housing. Those who lose under it are the poorer members of the community, who are liable to be priced out of their preferred areas and types of accommodation. Better income support measures for the poor would help but not cure the problem since if the poor could afford to pay more they would bid up the prices of popular types of housing and popular areas. The probable conclusion is that it is a good idea to have some but not all of the housing stock allocated by the market mechanism.

Administrative allocation will favour anybody the housing authorities favour strongly enough, and penalise any group the housing authorities do not favour. If allocation respects security of tenure of sitting tenants it is hard for the system to use housing efficiently, and if it does not it is likely to be greatly resented by tenants who feel they are being arbitrarily moved around or kicked out. However people with severe social problems are likely to find that public authorities are the only landlords willing to house them at all, and it is hard to see modern society doing without some administered housing. Except for specially favoured groups, administered housing is likely to be bad for mobility, and therefore for those who want to move. Thus while one would want there to be an administered housing sector, one would probably not favour making it a very big one.

The controlled market sector is good only for sitting tenants, who want to remain sitting in the same place and type of house; it is bad for landlords and practically everybody else. Because it is so bad for landlords it tends to shrink, whatever size one might think it ought to be.

2

The Facts about Housing in the UK

Before embarking on a discussion of various aspects of housing policy, this Chapter seeks to bring together some of the facts about housing in the UK. These are grouped under five headings:

Investment in housing
Changes in the number of dwellings and households
Changes in the condition of the housing stock
Changes in tenure
The finance of housing, both private and public sector.

INVESTMENT IN HOUSING

The UK has over the past decade spent a great deal on investing in housing. Table 2.1 shows figures for gross investment in housing at 1980 prices. This has varied between a maximum of £9.4b in 1979, and a minimum of £7.2b in 1981. In 1985 gross investment in dwellings was £8.5b at 1980 prices, very close to the average for 1975–85.

Not all this investment increased the stock of housing, since the houses already in existence are gradually wearing out. Capital consumption, *i.e.* wearing out, of housing is estimated as having grown from £3.3b in 1975 to £4.6b in 1985. On average over the decade, capital consumption was just under a half of gross investment in dwellings.

Net investment has varied between a maximum of £6.0b in 1976 and a minimum of £3.1b in 1981 and 1982; in 1985 net investment of £3.9b was a bit below the average for the decade. Net investment has been positive every year, so that the total stock of housing has been growing; this accounts for the growth of capital consumption.

17

Table 2.1: Investment in dwellings, 1975–85, at 1980 prices

£ billion

Year	Gross investment	Capital consumption	Net investment
1975	9.1	3.3	5.8
1976	9.4	3.4	6.0
1977	9.0	3.6	5.4
1978	9.0	3.7	5.4
1979	9.4	3.8	5.5
1980	8.7	4.0	4.7
1981	7.2	4.1	3.1
1982	7.3	4.2	3.1
1983	8.4	4.3	4.1
1984	8.9	4.4	4.4
1985	8.5	4.6	3.9
Mean 1975–85	8.6	3.9	4.7

Source: Blue Book on *United Kingdom national accounts*, 1986 edition, Tables 10.2, 11.2 and 11.4

Table 2.2: Private sector investment in dwellings, 1975–85, at 1980 prices

£ billion

Year	Gross investment	Capital consumption	Net investment
1975	5.5	2.2	3.3
1976	5.6	2.3	3.3
1977	5.6	2.4	3.2
1978	5.9	2.5	3.4
1979	6.5	2.6	3.9
1980	6.1	2.7	3.4
1981	5.5	2.8	2.6
1982	5.5	3.0	2.5
1983	6.1	3.1	2.9
1984	6.6	3.3	3.3
1985	6.5	3.4	3.0
Mean 1975–85	5.9	2.8	3.2

Source: Blue Book on *United Kingdom national accounts*, 1986 edition, Tables 10.2, 11.2 and 11.5. Capital consumption at 1980 prices is not available for private sector, and has been calculated by assuming that sector shares in total capital consumption at 1980 prices were similar to shares at current prices

Table 2.3: Public sector investment in dwellings, 1975–85, at 1980 prices

£ billion

Year	Gross investment	Capital consumption	Net investment
1975	3.6	1.1	2.4
1976	3.8	1.2	2.6
1977	3.4	1.2	2.2
1978	3.1	1.2	1.9
1979	2.8	1.2	1.6
1980	2.6	1.2	1.3
1981	1.7	1.2	0.5
1982	1.8	1.2	0.6
1983	2.3	1.2	1.7
1984	2.3	1.2	1.1
1985	2.0	1.2	0.8
Mean 1975–85	2.7	1.2	1.5

Source: Blue Book on *United Kingdom national accounts*, 1986 edition, Tables 10.2, 11.2 and 11.5. Capital consumption at 1980 prices is not available for public sector, and has been calculated by assuming that sector shares in total capital consumption at 1980 prices were the same as at current prices

The value of the 1980 net stock of dwellings at 1980 replacement cost was estimated as £248.5b, so the stock has been growing at about 1.9 per cent a year over the past decade.

These investment figures take no account of spending on home improvements carried out privately; in 1985 these were £3.4b at 1980 prices, including do-it-yourself materials and contractors' charges, but not putting any valuation on do-it-yourself labour.

It is of interest to see the corresponding facts for the private and public sectors separately. Table 2.2 shows private sector investment in dwellings. Private gross investment varied between £6.6b in 1984 and £5.5b in 1975 and 1981–2. In 1985 private gross investment in dwellings at £6.5b was above the average for the decade. Capital consumption has grown from £2.2b in 1975 to £3.4b in 1985; on average, it was just under half of gross investment. Net investment by the private sector varied between £3.9b in 1979 and £2.5b in 1982. These figures take no account of purchases of existing dwellings, including council houses bought by their tenants, but capital consumption in later years includes allowance for the wearing out of council houses sold to tenants in earlier years.

Table 2.3 shows investment in dwellings by the public sector.

Gross investment has varied between £3.8b in 1976 and £1.7b in 1981; in 1985 public sector investment in dwellings at £2.0b was well below the average for the decade. Capital consumption has been £1.1b or £1.2b throughout the decade. Net investment has varied between £2.6b in 1976 and £0.5b in 1981; in 1985 at £0.8b it was well below its average for the decade. Net investment has however been positive every year. The investment figures take no account of sales of council houses to tenants, but the capital consumption for later years does not include wearing out of houses sold in earlier years; this explains why capital consumption has been stable though net investment was positive.

Investment in housing may result in more houses or in improvements in the condition of existing homes. In fact both have occurred on a large scale, as shown in the next two sections.

CHANGES IN NUMBER OF DWELLINGS AND HOUSEHOLDS

One result of the considerable investment in housing has been a steady rise in the number of dwellings. To see the effects of this on housing conditions it is helpful to compare it with the number of households. Figure 2.1 shows the growth in the number of dwellings and households in Great Britain since 1951. Throughout the period dwellings have been growing faster than households. In the immediate post-war period there was an absolute housing shortage in the sense that the number of dwellings fell short of the number of households. By 1961, the figures were approximately equal, and from 1961 to 1974 there was a steady increase in the difference between them. Since 1974, the excess of dwellings over households has remained fairly stable at about one million. There is thus no overall shortage of housing, though there are shortages of housing of particular types or in particular areas. The overall availability of housing has clearly improved. Moreover, the great majority of dwellings in England and Wales in 1981, 94 per cent, were occupied by just one household. Recently, there has been a significant rise in the proportion of one- and two-person households and a decline in the proportion of households with three of more persons.

During this century, there has been a decline in the average size of households from 4.6 persons in 1901 to 2.5 by 1980. Factors affecting changes in household size are several and include fewer children, less stable marriages, greater longevity for the elderly and higher incomes allowing the young to set up separate households.

Figure 2.1: Households and dwellings, 1951–1986[a]

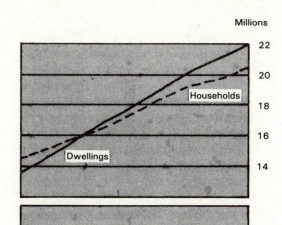

Note: a. Estimates of the stock in England have been revised and are based on the 1971 and 1981 Censuses. Estimates for Wales and Scotland prior to 1981 are based on the 1971 and earlier Censuses. Households figures from 1982 are projections from a 1981 base.
Source: Department of the Environment

The supply of housing also has effects on the number of households. People are more likely to set up separate households if separate dwellings are easier to acquire. Also, if dwellings become smaller they are less convenient to share, making households more likely to split and less likely to combine. Examples of splitting include children setting up house independently; examples of combining include elderly parents moving in with children.

The fall in the average size of households has had important implications for the demand for housing given the rise in population in the post-war period. While the population has increased by nearly 20 per cent, the fall in the average size of households has caused, in part, the growth in the number of households by over 30 per cent.

CHANGES IN CONDITION OF THE HOUSING STOCK

As well as leading to an increase in the number of dwellings, investment in housing has greatly improved the condition of the housing stock.

21

Table 2.4: Housing conditions in Great Britain, 1971–84 (availability of amenities)

	1971	1977	1979	1981	1982	1983	1984
With bath or shower (%)							
Sole use	88	93	95	96	97	97	97
Shared	3	3	3	2	2	1	1
None	9	4	3	2	2	1	1
With WC (%)							
Sole use inside the accommodation	85	92	93	95	96	97	97
None	1	–	–	–	–	–	–
With central heating (%)							
Night storage heating only	8	9	7	6	6	5	6
Other central heating only	26	41	47	52	54	58	60
Both kinds	1	1	–	–	–	–	–
Neither kind	65	49	45	41	40	36	34
Sample size[a] (=100%) (numbers)	11,852	11,815	11,312	11,855	10,202	9,908	9,671

Note: a. Smallest base of the three groups, for each year

Source: *General household survey*

Since 1951, there have been major improvements in the availability of basic housing amenities. For example, the proportion of households in Great Britain entirely without a fixed bath fell from 37.6 per cent in 1951 to just one per cent by 1984. Under all categories shown in Table 2.4, there have been remarkable improvements in availability between 1971 and 1984. During the period, the proportion of houses with central heating has increased enormously. According to the *General Household Survey*'s bedroom standard, there was also a tremendous decline in overcrowding.

Table 2.5 shows that the general improvement in housing standards in Great Britain between 1971 and 1984 applied to every tenure group. The availability of amenities has remained lowest in the private rented sector.

Repair, renovation and improvement

The report on the *English House Condition Survey 1981* estimated that £6.1b was spent in that year by householders on repairs, improvements, maintenance and decoration of the existing stock. If DIY labour were included this figure would be even larger. Updating this official figure of £6.1b in 1981, the Building Societies Association (1986) makes the guesstimate that over £10b is now being spent each year on home improvements. According to a major market research survey sponsored by the Building Societies Association in the spring of 1986, only 34 per cent of households need to raise extra loans to carry out improvements; this proportion rises to over 60 per cent, however, for those planning a full-scale conversion job. Installing new windows tops the list of improvements, with over a third of households having either put in double glazing or refitted their windows in the past five years. A quarter of households in the Building Societies Association survey had refitted their kitchens, one in five had tackled the bathroom, and nearly as many had completed rewiring in their homes.

For all past improvements, there is still concern about the present state of the housing stock. Indeed, appalling indictments of Britain's housing stock have been made recently. For example, the *Report of the Inquiry into British Housing* (1985) has claimed:

The scale of the national house condition problem is large, and the cost of remedying all defects is enormous. For example, 4.3 million homes or 24 per cent of housing stock of England in 1981

Table 2.5: Housing standards in Great Britain, 1971 and 1984 (by tenure)

| | Households lacking sole use of[a] | | | | With central heating (%) | | Sample size (=100%) (numbers) | |
| | Fixed bath/ shower (%) | | WC inside building (%) | | | | | |
	1971	1984	1971	1984	1971	1984	1971	1984
Owned outright	12	3	13	3	39	66	2,654	2,312
Owned with mortgage or loan	4	–	5	–	57	82	3,206	3,357
All owner-occupiers	7	1	9	1	49	75	5,850	5,669
Rented from local								
* authority/new town*	3	1	5	1	24	58	3,691	2,805
Rented privately unfurnished[b]	33	9	37	7	15	47	2,043	947
Rented privately furnished	58	29	57	29	17	36	320	217
All tenures	12	3	13	2	34	66	11,914	9,638

Notes: a. Excludes those living in caravans or houseboats; b. Includes those renting from a housing association, and those renting with a job or business.

Source: *General household survey, 1971 and 1984.*

was unfit, lacked a basic amenity or needed repairs costing more than £2,500. The total cost of remedying house condition problems in Britain as a whole in 1981 was, we estimate, about £35b.

The apparent enormity of the problem is placed in context when it is noted that total public expenditure on housing in 1981–2 was £3,100m. The most recent general report of the physical condition of housing stock was the *English House Condition Survey 1981*. It showed that of the 18.1 million dwellings in England, some 16.1 million or 89 per cent of the stock, were fit, had all basic amenities and did not require repairs costing more than £7,000. This represents a slight improvement compared with 1976, when the Housing Condition Survey showed 14.9 million dwellings, or 87 per cent of the stock, were fit. Using 1981 housing data, there appear to be some 1.1 million dwellings unfit and some 0.9 million lacking a basic amenity and one million requiring repairs of at least £7,000. Allowing for overlap between these categories, it seems unlikely that more than two million dwellings were in 'poor' condition in the early 1980s compared with 2.2 million in the mid-1970s.[1] This represents a slight improvement over the last decade.

Whether the situation is as bad as graphically described is debatable since it depends on the criteria employed. For example, since repairs and maintenance are made to houses in arrears it is not surprising that at any moment in time a substantial number of houses do need repair. The capital tied up in housing is considerable and thus an absolute figure attributable to depreciation is bound to be large. Two further points need to be made. First, the arrears at any moment in time should not be simply compared with the £3b of public expenditure. If this was all that was available, the problems would get worse. In fact, there was, in 1981, a further £6b of private maintenance and improvement expenditure. Second, it is not clear how much of any required spending was in respect of dwellings which in declining areas could already be unoccupied, and how much was in respect of dwellings in other areas which are due to be displaced in the near future by newly built ones. It is these two factors which mean that the problem could be expected to shrink even in the absence of extra public expenditure, though an increase could help it shrink faster.

Suggestions that the state of the nation's housing stock is declining and that government is not spending enough on improvement and refurbishment seem overstated. Clearly, pockets or housing are declining, especially in some inner-city areas, but it would be wrong

25

to think this was widespread. Moreover, public expenditure remains considerable. In 1984, 87,000 local authority and new town dwellings in England were renovated, a higher figure than in any year since 1973. During this decade, over 450,000 local authority and new town dwellings have been renovated. In addition, the government provides finance to local authorities through the Urban Housing Renewal Unit, now renamed Estate Action, for special schemes to renovate estates and bring empty flats back into use for the homeless. Finance for these schemes has risen from £50m in 1985–6 to £75m in 1986–7.

CHANGES IN TENURE

The twentieth century has seen an extremely rapid change in the structure of tenures of housing in the UK. There has been a rapid increase in the percentage of houses which are owner-occupied, or rented from the public sector, and a rapid decline in the private rented sector. The percentages are shown in Table 2.6, and the actual numbers in each sector in Figure 2.2.

In 1914, only ten per cent of dwellings in England and Wales were owner-occupied. By 1939, this proportion had increased to 32 per cent, by 1960 to 44 per cent, and by 1971 to 50 per cent. The proportion increased to 53 per cent in 1976, to 57 per cent in 1981 and 60 per cent by 1983. At the end of 1986, 63 per cent of dwellings were owner-occupied.

Table 2.6: Distribution of dwellings in the UK, 1914–85 (by tenure)

	Owner-occupied	Rented from Public authorities or new town corporations	Rented from private owners[a]
1914	10	—	90
1950	29	18	53
1960	42	26	32
1975	53	31	16
1985	63	28	9

Note: a. Includes housing associations and dwellings rated with farm or business premises.
Source: *Housing and construction statistics*.

Figure 2.2: Stock of dwellings in the UK, 1951–85 (by tenure)[a]

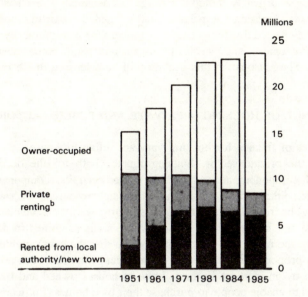

Note: a. Estimates of the stock in England have been revised and are based on the 1971 and 1981 censuses. Estimates for Wales and Scotland prior to 1981 are based on the 1971 and earlier censuses. Estimates for Northern Ireland are largely based on rate returns; b. Includes housing associations and dwellings rented with farm or business premises, and those occupied by virtue of employment.
Source: Department of the Environment; Department of the Environment, Northern Ireland.

There has also been a rise in the percentage of dwellings provided for rent by the public sector, mainly local authorities and new town corporations. This started in the inter-war years, and expanded greatly in the post-war period. By the late 1970s public sector housing accounted for 33 per cent of the stock of dwellings. In the last decade this percentage has started to decline, through a reduction in local authority house building and sales of council houses to sitting tenants; but in 1985 the public sector still provided 28 per cent of the housing stock, and considerably more in some areas.

These changes have been accompanied by a continual decline in the private rented sector. In 1914 90 per cent of housing was privately rented, and this share has fallen to under ten per cent at the end of 1986. Not merely the percentage share but the actual number of privately rented dwellings has declined. There has been very little building for private renting, and the existing stock of privately rented dwellings has been reduced by demolition, by sales to sitting tenants, and by sales for owner-occupation when vacancies arose. The reasons for this decline will be discussed in Chapter 5.

FINANCE OF HOUSING — PRIVATE AND PUBLIC SECTOR

Sources of finance for housing vary according to tenure.

In the private sector, housing finance both for the declining rented component and expanding owner-occupied component is derived either from the private financial resources of owners (equity) or from financial institutions (mortgages). In Chapter 3, the growth and importance of these institutions is examined in detail. For the moment, it is important to stress that the considerable rise in the proportion of houses in owner-occupation has been accompanied by an increase in the volume of houses owned and finance needed to enable people to purchase their own homes. For example, the value of dwellings owned by the personal sector doubled from 23 per cent of personal net wealth in 1957 to nearly 50 per cent in 1985. Loans for house purchase also increased in importance in the personal sector balance sheet and at the end of 1985 represented over ten per cent of net wealth compared with five per cent in 1955. Similar trends are evident in the balance sheets of financial institutions. At the end of 1985, building societies accounted for over 20 per cent of all personal sector financial assets compared to five per cent in 1955. Moreover, the building societies' market share of the liquid savings market, excluding pensions, life assurance and marketable securities, has grown from only ten per cent in 1950 to nearly 50 per cent at the end of 1985.

Chapter 6 emphasises the relevance of housing finance and taxation. In particular, public 'subsidies' are arguably granted to owner-occupiers in the form of tax relief on mortgage interest. There is a considerable debate as to whether this constitutes a subsidy and also a conceptual difficulty in the estimation of this concession. The real concession arises from owner-occupiers getting services out of gross income, while tenants have to pay their rent out of net income. Of course, not all owner-occupiers' houses are mortgaged and not all mortgages are anywhere near 100 per cent. That income tax relief is frequently regarded as a subsidy arises from government tax distortions in 1963 with the abolition of Schedule A tax and in 1974 with the cavalier abolition of tax relief on private borrowing, save owner-occupied house purchase, whereby tax relief on mortgage interest is permitted on loans up to £30,000 maximum. Moreover, tax relief is largely funded by the building societies on behalf of their investors. The controversy centres on whether tax relief on mortgage interest is a subsidy or not but, in any event, it cannot be directly compared with the average subsidy granted to local authority tenants as a result of other complications.

In the public sector, housing finance is derived from a number of sources. First, central government provides subsidies to local authorities. In recent years the value of these subsidies has fallen quite dramatically from £2,029m in 1980–1 to £656m in 1983–4. Second, housing finance can be derived from ratepayers through General Rate Funds. Such contributions have increased slightly from £523m in 1980–1 to £656m in 1983–4. Third, authorities may spend each year up to a specified proportion of capital receipts accruing from various sources.[2] Receipts from council house sales are available to finance new building, *etc.* Receipts not used are invested to produce income which in turn is a credit to the housing revenue account. Finally, rent income provides a further source of finance for public sector housing. Although these sources are all significant the exact relative combinations differ considerably between authorities, and over time. The reasons for this will be examined in Chapter 4.

Net public expenditure provision for housing includes current spending on subsidies (this excludes spending on housing benefit, which comes under the social security programme) and on administration, plus capital expenditure on new public sector building, refurbishment of the public sector stock, support for owner-occupation through low-cost home ownership initiatives, and improvement grants in the private sector. The total does not include capital receipts, which are shown separately in Figure 2.3. These accrue mainly from the sale of council houses. Figure 2.3 shows housing expenditure in cash terms, *i.e.* not adjusted for price infla-tion, since this is now the normal way that government expenditure programmes are primarily planned.

During the 1980s, the public expenditure programme, in cash terms, declined from its peak of £5.5b in 1979–80 and 1980–1 to around £4b in 1981–2 and thereafter. Capital receipts rose steadily to £2.5b in 1983–4 but declined thereafter. Housing benefit — rent rebates and allowances — which stood at £298m in 1979–80

Figure 2.3: Public expenditure on housing in the UK, 1980–7

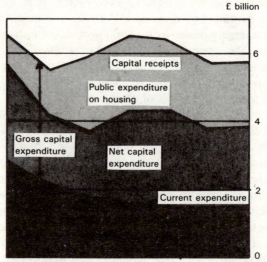

Source: *The Government's expenditure plans*, Cmnd 9702, HM Treasury, HMSO.

increased to £2.5b in 1983–4 and £2.8b in 1985–6. At the same time, the number of recipients increased from 1.4 million in 1979–80 to 4.7 million in 1985–6.

All public sector housing finance has been strictly controlled during the past decade as part of central government overall policies on public sector expenditure. Housing controls cover capital spending on council house building, modernisation and repair, slum clearance, renovation grants and loans to individual householders, and loans to housing associations. Local authority housing finance is extremely complex; not only are the sources several but the uses are diverse, relating not only to council housing but specific financial provision for houses to let and ownership in the private sector. By law, all local authorities maintain a Housing Revenue Account (HRA) and it is in this account that all expenditure and income relating to the provision and management of council housing are recorded. Until 1980, HRAs had to balance (with an allowance for a 'working' balance), but since 1980 an HRA must not be in 'deficit', *i.e.* there is a scope for 'profit'. However, while some authorities may set rents to budget for a surplus, others may keep rents low and, if necessary, actually provide a subsidy from the General Rate Fund to avoid a deficit on their HRA. It is clear that the financial decisions of any one local housing authority depend upon its own financial position, its block grant from central government, its overall expenditure limits ('rate capping'), and also its own view as to what rents it should collect. Upon these issues depend what local authorities decide to spend on house building, repair and maintenance, management and administration.[3] For some authorities, it will mean the potential for transferring any 'profit' on their HRA to their General Rate Fund.

The large number and diversity of the public and quasi-public bodies responsible for housing expenditure and the multifarious expenditure programmes distinguish housing in a number of ways from other social programmes where ownership and resources planning is more centralised and policies are, to a greater extent, determined nationally. These special characteristics of housing are described below.

Housing expenditure is directed towards both the public and private sectors. Although the major proportion of expenditure is directed towards public-sector housing, certain expenditure is common to all tenures. Moreover, the allocative criterion of need, so firmly entrenched in the provision of health care, is only vaguely applied for housing resource allocation and is frequently decentralised both in theory and practice. For example, local authorities are not only responsible for their own housing programmes but have great discretion in such matters as the actual allocation of council housing and allocation of improvement grants in the private sector, as well as exercising great control over planning matters.

Policy is decentralised among 400 housing authorities. Local authorities have some freedom to form their own priorities, taking into account government policies. A Housing Investment Programme (HIP) is compiled each year and submitted to the Department of the Environment. The amount of capital expenditure an individual authority can undertake in any year depends on the combination of a prescribed proportion, which is at present 20 per cent for most housing receipts in the relevant year, and accumulated capital receipts.[4] An HIP allocation is made to that authority each year by the Department. Each authority can exercise considerable freedom in both house-building and improvement programmes and the criteria to be applied locally for the allocation of council housing. Thus local authorities not only decide on those individuals and families eligible for housing allocation but are able to determine the relative priority to be extended to different categories. In addition, local authorities are sovereign in the formulation of contracts specifying the obligations of tenants in respect of repairs and maintenance as well as the form and method of rent payment. It has only been in the last ten years that government has imposed limits and controls on local housing expenditure. Before 1976, much expenditure emanating from central government was open-ended, with Exchequer subsidies provided to local authorities.

In view of the broad scope of housing expenditure, there are considerable problems of analysis, for it is extremely difficult to relate the scale and form of expenditure to the scale of housing services provided. Unlike health and education, where the state in its widest sense owns a high proportion of resources, the ownership and allocation of the housing stock is mainly limited to housing authorities. In addition to these functions, these are also responsible for financing, either as agents of the central government or as principals, housing activities in both private and public tenures. Moreover, the Inland Revenue has responsibilities for tax relief on housing advances to owner-occupiers, the DHSS for supplementary benefit and local councils for the administration and provision of housing benefit to council tenants and to private tenants.

According to *The Government's Expenditure Plans 1986–7 to 1988–9* (HMSO, 1986) gross provision for public expenditure on housing in 1986–7 (gross capital plus current) is estimated at £4,350m. After taking account of capital receipts of £1.6b, largely flowing from the Right-to-Buy scheme for public-sector tenants, the net programme will total £2,750m. No attempt has been made to estimate the concessions to owner-occupiers based on the non-taxation of imputed income, the value of the fiscal advantage to building societies nor subsidy to unfurnished tenancies enforced on private landlords. In 1986–7, HIP allocations are expected to total £1,465m, while housing benefit expenditure is projected at well over £3,000m per annum for the remainder of this decade. Such enormous state subsidy was neither intended nor anticipated in past decades; its growth has been accelerated by a combination of high inflation and high unemployment.

NOTES

1. In 1976, if 14.9 million fit dwellings were 87 per cent of the stock, the 13 per cent unfit were about 2.2 million. In 1981, if 16.1 million fit dwellings were 89 per cent of the stock, the 11 per cent unfit were two million.

2. The prescribed proportions applying to capital receipts accruing to local authorities in England are, in the case of receipts arising on housing services:

(i) 30 per cent of receipts from sales of dwellings held under Part II of the Housing Act 1985 and normally let or available for letting, which are disposed of on a shared ownership lease or for homesteading;

(ii) 20 per cent of receipts from other sales of dwellings (including receipts arising from payments of principal on sums left outstanding on the security of dwellings) held under Part II of the Housing Act 1985 and normally let or available for letting;

(iii) 20 per cent of repayments of principal on and redemption of advances for house purchase and improvement in the private sector;

(iv) 20 per cent of repayments of principal on service charges left outstanding by virtue of regulations under sections 450A and 450B of the Housing Act 1985;

(v) 1 per cent of repayments of loans to housing associations (including Housing Association Grant);

(vi) 100 per cent of receipts from sales of dwellings not normally let or available for letting. This category includes receipts: from homes built for sale by the authority; from homes bought in by the authority for resale; and from homes built by private developers under licence on the authority's land (provided that an appropriate form of licence is used); and

(vii) 30 per cent of receipts from the sale of undeveloped housing land.

3. Although a housing authority's overall finances might govern the amount to be made available to the HRA, house building is limited more often than not by the government-imposed capital spending allocation.

4. In theory, an authority can aggregate all its block allocation (including Housing Investment Programme) and capital receipts (properties) and use them for whatever purpose it determines. Accordingly, the HIP allocation could be used to provide a swimming pool!

Part II

Housing Tenure and Finance

3

Owner-occupation: Choice and Finance

PREFERENCES FOR OWNER-OCCUPATION

We have seen in Chapter 2 that in the course of the present century owner-occupation has grown from being a form of tenure enjoyed by only 10 per cent of households in 1910 to the typical form of tenure, with 63 per cent in 1986.

One important reason for this must be that people who can afford it prefer to own their own home. In 1910 only a small minority were well enough off to be able to afford to be owner-occupiers. The very large increase in real incomes since then has meant that people on average incomes or below can in many cases now afford what only the better off could attain in previous generations.

This view, that the rise in owner-occupation reflects people being able to satisfy their preferences as real incomes rise, can be supported by two pieces of evidence. First, people can be asked to state their preferences. According to a recent survey by the Building Societies Association (1986) no less than 95 per cent of owner-occupiers were 'quite satisfied' or 'very satisfied' with their housing. Only 38 per cent of existing council tenants were satisfied with their housing and the desire of tenants to purchase their homes has apparently not declined. Among all council tenants, 25 per cent thought it likely they would buy their own homes.

Most people do, however, have strong views about the type of tenure they want. The survey found that owner-occupation is the ideal. This preference is particularly strong among younger people: 65 per cent of those aged 16 to 19 wanted to have their own property in two years' time; this was true of 80 per cent among those aged 20 to 24 and 88 per cent of those 25 to 34.

By that age, of course, many have achieved their objective. While

37

63 per cent is the overall level of owner-occupation in Britain, 72 per cent of those in the 25 to 34 age group already own or are in the process of buying their home, a share exceeded only by the 74 per cent in the 35 to 54 age group.

This is one factor which quite distinguishes Britain from other countries, especially New Zealand, the United States and Australia, all of which have higher overall levels of owner-occupation. A far greater proportion of Britons in their 20s are buying their homes.

There has not been a year since 1945 when the relative share of owner-occupied housing stock has not increased. In effect, owner-occupation has been moving 'down market' embracing an increasing number of households.

A second piece of evidence is a consideration of the forms of tenure in other industrial countries. The details of institutional arrangements and tax rules concerning houses vary considerably between countries. Any characteristics of tenure patterns which are at all widespread are thus not likely to result from special institutional arrangements or tax rules. In fact owner-occupation is the prevailing tenure form in most industrial countries. With 63 per cent of houses in owner-occupation, Britain compares almost equally with Canada (62 per cent), United States (65 per cent), Japan (62 per cent) and Belgium (61 per cent).

It is however quite possible that the extent of the shift to owner-occupation may have been affected by factors other than the general rise in real living standards. One such factor is the lack of sufficient rented accommodation. If it is hard to obtain rented accommodation, this makes it more likely that people who can afford it will become owner-occupiers. Chapter 4 will consider whether the public housing sector has been run in such a way as to produce shortages of rented housing. Chapter 5 will consider whether policies towards the private rented sector may have produced a shortage of houses to rent privately. It will in fact be argued that the rented sector has been run in such a way as to tend to drive people towards becoming owner-occupiers.

It is also possible that owner-occupation may have been encouraged by the fiscal system. Chapter 6 will look at the system of taxation and its effects on the incentive to become an owner-occupier.

FINANCE FOR OWNER-OCCUPATION

When discussing whether anybody can afford to be an owner-occupier, it should be remembered that it is not necessary to be able to afford to buy a house outright from previous savings. Some households may be able to afford this; for example one can become a house owner by inheritance. One might expect that in a society in which 63 per cent of dwellings are owner-occupied, a large proportion of the population would one day inherit a house. One reason why this has not happened as yet is that a large proportion of owner-occupiers are at present quite young. It may never happen at all in some cases, since it is possible for the elderly whose families are grown up to sell their houses and consume the proceeds, while living with children or other relatives or in sheltered accommodation. It is also possible to sell one's house, retaining a life tenancy. However some people do inherit houses, while others are given or lent money towards house purchase by parents or other relations intending to live with them.

In any case, the typical householder in modern society can expect to live to an age at which it is likely that their children will already have started households of their own, which will usually involve buying a house. An inherited house is unlikely to be in the right place, or of the right type, for the heirs to wish to live in it. Inherited houses are usually sold, but the money will enable the heirs to afford a better house for themselves if they are already house owners, or to afford to become house owners themselves if they have not done so already.

The typical situation of the first-time house buyer, however, is that they do not buy a house outright from previously saved capital, but provide only a part of the total cost, often as little as five or ten per cent, and borrow the remainder on mortgage.

Sources of mortgage finance

The main providers of mortgage finance for house purchase are building societies, commercial banks, local authorities and insurance companies. Of these institutional sources building societies are by far the most important, responsible in most years for about 80 per cent of home advances. Of considerably less importance than a decade ago are local authorities, whose mortgages are today a very small percentage. Insurance companies provide finance for home purchase through the sale of endowment policies while local authorities are empowered to make loans, in theory at least, to those on low earnings and on property of lower quality which building societies might regard as poor collateral. Commercial banks are the most recent entrant into the mortgage market. A decade ago they would grant only 'bridging' loans, *i.e.* loans to allow the purchase of a new house before sale of one's present house is completed. Today they have established themselves as important lenders, even if a little erratically from one year to another, as Table 3.1 shows. To some extent, their entry has sent building societies 'down market' in their own lending. Nevertheless, building societies remain the major source of loans for house purchase.

Although the total assets of the building societies have continued to grow rapidly over the past decade, the actual number of building societies has continued to decline steadily. At the turn of the century there were over 2,000 societies but by 1930 the number had fallen to about 1,000. In the post-war period the number of societies continued to decline to just over 700 in 1960 and at the end of 1986 there were about 150 societies. This process of acquisition, merger and takeover has resulted in a heavy industrial concentration within the movement by the largest societies. The five largest building societies now account for about 60 per cent of total assets and the largest 20 for approximately 90 per cent of total assets. At the other end of the scale, there are a great number of localised societies with very small total assets.

From modest beginnings, building societies are now one of the largest groups of financial institutions in terms of size of asset holdings. In their involvement in the personal sector, they are by far the most important intermediary, for no other institution displays such a high level of interdependence within this one sector. Building societies have no equity capital as such, but over 95 per cent of their liabilities are to the personal sector in shares and deposits constituting short-term placements withdrawable on demand or at fairly short notice. Share certificates confer membership on the holder and are issued on demand to shareholders, who receive a fluctuating interest rate on these shares rather than a profit dividend. Depository funds, however, confer creditor status on their holders and in view of this they rank before shareholders in the event of dissolution. Usually the interest coupon is a quarter per cent below the share rate.

These liabilities enable building societies to make house purchase advances. Over 80 per cent of the combined assets of societies are

Table 3.1: Net mortgage advances in the UK, 1971–84 (by source)[a]

£ million

	Building societies	Local authorities	Insurance companies and pension funds	Banks[b]	Other sources	All sources
Net advances						
1971	1,600	107	17	90	12	1,826
1976	3,618	67	45	80	60	3,870
1980	5,722	454	264	593	300	7,333
1981	6,331	271	88	2,447	353	9,490
1982	8,147	555	6	5,078	353	14,139
1983	10,928	−344	124	3,639	22	14,369
1984	14,572	−209	212	2,314	−85	16,804
Amount outstanding at 31 December 1984	82,385	3,915	2,544	17,309	1,515	107,668

Note: a. Gross advances less repayments of principal; b. Includes Trustee Savings Bank.
Source: *Financial statistics*, Central Statistical Office

advanced to the personal sector through mortgage advances exclusively for house purchase, while the balance is held in government securities and liquid assets.

Regulations arising from trustee status require the building societies to satisfy two basic ratios in their portfolio structure. A reserve ratio on a sliding scale relates general reserves to total assets, and a liquidity ratio is of greater importance and has a threefold rationale: to cover withdrawals of shares and deposits, to act as a cushion for increases in advances, and finally to cover seasonal elements in the inflows and outflows of funds.

Loans for house purchase

Lending institutions, and building societies in particular, have been critical to the growth of owner-occupation in Britain, particularly as many people seek to become owner-occupiers at a comparatively young age. Such households have only a limited amount of savings and as their incomes are modest and have not yet reached a peak, they are only able to afford cheaper houses. As their income increases, however, people move up market both with and without the support of housing finance. It is the housing finance system which provides finance to assist young first-time buyers who have limited capital and income. The amount that a society will lend to a prospective purchaser will depend on the value of the property and on the ability of the borrower to repay the loan. Mortgages advanced by building societies are either of the annuity type or of the endowment type as life assurance policies have become more common.

Under the annuity mortgage equal or 'level' monthly payments are made throughout the term (normally 25 years) of the mortgage, each payment consisting partly of a repayment of the loan and partly of interest on the amount of the loan still outstanding. At the outset of the loan, the bulk of each payment consists of interest which can be fully offset against personal income tax. This means that the net cost of the annuity method of repayment is low in the early years of the mortgage, increasing in net fixed terms as more capital is repaid and a smaller proportion of each successive payment, on a yearly basis, comprises interest. Not only is this method beneficial to building societies since capital repayments start being made, albeit very low in the early years, which can be re-lent to new mortgagors, but it is also the cheapest form of loan involving capital and interest repayments, so far as basic rate taxpayers are concerned.

Endowment mortgages consist of a loan, from either a building society or a life assurance company, secured by an endowment life assurance policy. During the term of the policy all payments are premiums on the policy, and interest on the loan. At the maturity of the policy, the proceeds are used to repay the loan. The main attraction of such mortgages is that while the annual net-of-tax cost tends to be higher in the early period of the mortgage, the overall net cost, bearing in mind the rising net cost of an annuity mortgage towards maturity, is lower, together with the prospect of bonuses arising from a with-profits policy, although it should be recognised that unless a person actually wants life assurance they may well be better off with an annuity mortgage. Tax relief on premiums is now no longer available on new endowment policies.

The rate of interest applied to mortgage advances is variable and will fluctuate according to, but less frequently than, prevailing money market rates. Until the collapse of the building societies' interest rate cartel in 1984, the majority of building societies conformed to the recommendations of the Building Societies Association in their lending and borrowing rates. At any one moment, the very great majority of mortgages bore the same borrowing rate. The societies' new emphasis on paying a competitive rate of interest rather than increasing the level of service has been reflected in the decline of new branches opened in 1985 and 1986. New financial pressures on societies are forcing their managements to pay more attention to the cost of running societies.

Although membership of a society is the primary allocator of mortgages to potential borrowers, the lending rate, the credit-worthiness of the borrower and the collateral value of the house are important considerations. The credit-worthiness of an applicant embraces personal status with respect to income, age and occupation and the normal requirement is that the weekly income is greater than monthly mortgage repayments or that the advance should not normally exceed two and one-half times the annual income, with certain allowances for the income of the mortgagor's spouse. Since the house provides the collateral for a loan, the age, type and condition are important factors in the determination of mortgage advances. Societies are not legally able to make an advance for more than the value of the dwelling as assessed by valuation. Some 75 per cent of first-time buyers obtain an advance for 80 per cent or more of valuation. Given additional security, such as a guarantee provided by an insurance policy in return for a single premium, the percentage may be increased to 95 or even 100 per cent.

DRAWBACKS WITH OWNER-OCCUPATION

It is quite probable that with further growth of real incomes, owner-occupation will increase its share of the housing market still further. To assess how much scope there is for further growth, it is necessary to consider its disadvantages; these are sufficient to make it very unlikely that owner-occupation will ever become universal.

First, owners are fully responsible for the maintenance and repairs to their houses. This is partly a matter of routine maintenance, including things like regular repainting, remedying minor damage to windows, guttering and roofs, periodic rewiring, and so on (any householder will confirm that the list is almost endless). It is also a matter of coping with exceptional incidents like subsidence of foundations, cracks in walls, severe storm damage, infestation by woodworm, and various forms of wet and dry rot. This maintenance work demands frequent attention and expenditure. If maintenance is not done regularly and competently the various exceptional problems all become more likely to occur, and the damage costs more to remedy the longer it is neglected. Because of this sensitivity to the standard of routine maintenance, insurance against the cost of major repairs is always expensive and often not available. In any case insurance covers only the cost and not the trouble of repairs.

All this means that it is only prudent to become a house owner if you are mentally and physically fit to cope with the trouble and are financially fit to cope with the costs of maintenance and repairs. Owner-occupation is thus not really suitable for several classes of people. These include the very poor, who are liable to have severe problems in raising the finance for essential maintenance work. They also include the elderly, the disabled, and the merely incompetent who are not able to organise the necessary inspection and remedial work to keep their homes in good condition. It is highly probable that a lot of owner-occupied housing is at present deteriorating because the owners are too poor to pay for necessary maintenance, or unable through old age or other causes to organise it.

A second group for whom renting is really more sensible is those whose occupations involve frequent changes of address. Buying and selling houses is expensive of time and money; this is only a small proportion of the benefits gained from a house over a long period, but becomes disproportionate for a stay of a few months. It is very risky to cut down on the time and expense of buying a house, since if one buys a house which turns out to have unexpected defects, selling it again may only be possible with loss or delay (or both if major building work is required). For people whose careers require mobility, renting a house or flat makes more sense than buying, since somebody else is taking on the trouble and risks of maintenance — the cost of course will still fall on the tenant since it should be reflected in the rent charged.

Since it seems very probable that the poor will always be with us, and it is quite certain that a lot of people will be old, incompetent or highly mobile, it seems pretty clear that owner-occupation will not suit everybody. There will always be a need for rented housing to be available.

4

Public-sector Housing: Finance and Allocation

AIMS OF PUBLIC-SECTOR HOUSING

The principal aim of public-sector housing is the provision of housing for those who have none, and of satisfactory housing for those whose existing housing is sub-standard. Public-sector housing includes bodies and authorities in receipt of direct financial provision from the central government for the purposes of building, acquiring and maintaining houses for rent. By far the most important are the local authorities and housing associations. A major aspect of local government policy involves the renovation of the existing housing stock, while building by local authorities and housing associations has been increasingly directed towards provision for the elderly, the disabled and those whose needs may be met only with difficulty in the private sector.

There are some further aims: these have tended to cause the amount of housing provided by the public sector to expand beyond the minimum level needed to satisfy the main aim. These include:

1 Elimination of private renting, on moral grounds and/or because it is believed that conditions in private renting cannot be made satisfactory;

2 'Town planning' considerations, based on a belief that public ownership of whole areas can improve their layout and environmental quality;

3 A desire to have a social mix in public housing, and avoid a 'ghetto effect' of housing only people with social problems;

4 Inertia effects: if people are allocated public-sector housing under the first aim, and there is a reluctance to disturb sitting tenants, satisfaction of the primary aim when further cases of no

46

or bad housing arise mean that the primary aim can be satisfied only by extending the public sector;

5 Pure empire building by those who administer public sector housing.

There is evident dissatisfaction over public sector housing, partly because of tenants' preference for other tenure, and partly because of the poor condition of some of the stock of public sector housing. Capital expenditure on repair and renovation in 1985-6 was £1.2b, two and one-half times as much as in 1978–9, when such expenditure was £480m. Renovation is now accounting for 40 per cent of council house capital expenditure compared with 21 per cent in 1978–9. Some attempt has been made to improve council house estates through the urban and community programmes and grants are available to private-sector building firms to acquire and refurbish estates and sell them for rent or purchase. In total councils are spending over £2b each year on renovation, maintenance and repair of their housing stock on current and capital accounts.

LOCAL AUTHORITY HOUSING FINANCE

Although local authorities are generally thought of as being mainly concerned with the provision of housing for rent, this is only one, albeit important, part of their multifarious responsibilities which encompass all tenure groups from the provision of housing advances, to housing benefit subsidies, and slum clearance and renovation programmes. Total capital expenditure on local authority housing has increased from £533m in 1966–7 to £1,733m in 1975–6 (its historic peak, after adjustment is made for inflation) and to £2b in 1985–6. During the same period, however, the proportion of housing revenue expenditure financed from rents and service charges after deducting rebates, has decreased considerably. This proportion was 74 per cent in 1966–7, 44 per cent in 1975–6, and 34 per cent in 1985–6.

Separate financial budgets and prescribed expenditure limitations invariably pose problems for any local authority which seeks to implement an overall housing strategy. This can result in a piecemeal and somewhat compartmentalised approach to housing issues, with the ownership and allocation of council houses as a primary function and all other functions of secondary importance.

To describe the limitations imposed on the housing functions of local authorities, it is convenient to classify these activities according to their actual financing. Broadly, there are three main ways by which activities can be financed, owing more to historical development than any functional rationale; they create a number of anomalies due to the inherent inflexibility of such administrative classifications under separate accounts.

(i) Self-financing activities

These activities include the provision of private mortgages and advances to housing associations. Each account must be wholly self-financing with no subsidy element from the rate fund, housing revenue account or direct government finance.

(ii) Rate fund finance

This account may finance a number of activities in the private sector and housing associations, but local authorities may use rate fund finance only in conjunction with certain and separate financial contributions from central government. Such activities include housing advice centres, general financial support to housing associations, improvement grants, general improvement areas (GIAs), housing action areas(HAAs) and slum clearance programmes.

(iii) Housing revenue finance

The housing revenue account may be used only in connection with the building, management and financing of local authorities' own stock of housing and expenditure incurred in the purchase of existing houses as additional stock. In some sense, this account may be considered a consolidated account, because all rental income, government subsidies, rate fund subsidies, and interest receipts[1] from the sale of local authority housing are credited to the account and all expenses relating to acquisition, maintenance, supervision and administration of the authority's stock of houses charged against it.

The distinct, but interlocking relationship, between the rate fund and the housing revenue account is important. First, certain contributions may be made from the rate fund to the housing revenue account and vice versa. Second, although the accounts are maintained quite separately, all loans raised by a local authority, irrespective of the purpose for which they have been raised, e.g. housing, leisure or industrial development, must be pooled and an average rate of interest apportioned and charged against the general rate fund account or housing revenue account according to the functional purpose of the loan.

Table 4.1: Summary of local authority Housing Revenue
Accounts, 1984–5 (estimated)

£ million

Expenditure		Income	
Supervision and management	912	Gross rent	3,413
Repairs and maintenance	1,254	Interest	502
Interest payments	2,493	Other rents and income	360
Debt repayments	359	Rate fund contribution-rate	
Capital met from revenue	53	rebate administration	19
Other expenditure	153		
Change in balances	−89	Total income	4,294
Total expenditure	5,135		
HRA deficit	841		
Financed from Exchequer			
subsidy	393		
Voluntary rate fund			
contributions	486		
Transfers to the general			
rate fund	−38		

Source *The Government's expenditure plans 1986–7 to 1988–9*, Vol. 2,
Cmnd. 9702, HM Treasury, HMSO, January 1986

Since housing represents a major capital expenditure for all district and city councils, the ongoing building programmes, following local authority reorganisation in 1974, have considerably increased the outstanding debt of local authorities.

Until 1980, local authorities were required to balance their HRAs subject to a permissible working balance (see p. 30). The main sources of income to the HRAs are council tenant rents, rents for garages, subsidies from the Exchequer, rate fund contributions, and interest earned on the invested receipts derived from the sale of council houses.

The main expenditure items of an HRA, shown in Table 4.1, are debt charges comprising interest and repayment of capital, housing management and maintenance and repair costs. Although the actual amount of Exchequer subsidy payable to a local authority is outside the control of councils, they are free to budget for surpluses or deficits (subject to contributions from the general rate fund) and set rents accordingly. Between 1978–9 and 1982–3, the government encouraged authorities to increase the real level of rents so that tenants would meet a larger part of their own housing costs.

Table 4.2: Council house subsidies, 1978–85[a]

£ million

Year	Subsidies to housing	Rent rebates	Total
1978–9	1,291	207	1,498
1979–80	1,696	238	1,934
1980–1	1,968	317	2,285
1981–2	1,426	490	1,916
1982–3	1,110	922	2,032
1983–4	779	1,756	2,535
1984–5		1,900	

Note: a. Central government grants to local authorities.
Source: *The Government's expenditure plans* (White Papers for relevant years), HM Treasury, HMSO

Subsidies to council housing have therefore fallen sharply, but subsidies to council tenants because of housing benefit have not. Table 4.2 shows the changes in council housing subsidies in recent years. Two-thirds of local authority tenants receive rent or income support, and it is estimated that benefit payments represent about 55 per cent of rents collected by local authorities.

Decisions facing a local authority, however, concerning the number of houses to build, improve, or the rents to charge, have little to do with the real influences on the HRAs, because financial decisions are heavily dependent on central government block grants and possible expenditure limits due to rate capping. Also local authorities may or may not decide to transfer 'surplus' or 'profit' to their general rate fund as they may now legally do. Thus these decisions depend far more on the political nature and composition of the council than on financial or housing need. Not surprisingly, the state of local HRAs varies widely. While some councils make transfers to the general rate fund, others, especially inner city authorities, make massive transfers from their rate funds to their HRA. As an example, in 1983–4 in England, £29m was transferred from HRAs to general rate funds, while £497m was transferred from general rate funds to HRAs. The net effect was a net transfer from general rate funds to HRAs of £468m.

RENTS AND SUBSIDIES

In an economic analysis of public sector housing, the issue of public housing subsidy is treated as being a direct payment or price subsidy from central or local government to an individual. Subsidising housing by state provision below market-rent housing may be justified, or indeed criticised, in two main ways. First, the existence of some externality, *e.g.* health, income distribution, may cause society's demand curve for housing to lie above the market demand curve. Because of considerations of this sort, governments may well have determined, as a matter of policy or regulation, some minimum desired level of housing consumption *per capita*. Such a level of consumption of housing may be achieved by raising incomes, through income transfers or redistribution, so indirectly raising housing consumption. An alternative approach to achieve similar goals of housing consumption is by distorting housing service prices, *i.e.* by imposing maximum or nominal prices.

An economic subsidy may normally be defined as the difference between a market-clearing rent and an actual rent. Using this measure in the case of council housing is impracticable due to the absence of a free market in which the market-clearing level of rents can be observed. Instead, one might adopt an accountancy measure, which is the difference between replacement cost and actual rent. However, the costs of land make this difficult also. An alternative accountancy measure is the difference between historical costs and actual rents. It is the latter definition which is used in discussion of council housing rents. Of course, some tenants are also in receipt of further financial benefits in the form of housing benefit if their income is deemed to be too low to meet in full the rent applied to their council home.

Council housing allocated according to 'need' has been actively encouraged by governments through loans and grants to local authorities and housing associations. The nature and forms of subsidy have changed frequently but their common effect has been to reduce the full cost burden of house-building falling on local authorities. Rents have traditionally been set at below market price under a system known as rent pooling. Thus the rent of a house is not determined by the actual original or replacement cost of the house, but after all housing costs of an authority have been pooled; pooling also takes into account all subsidies, and any Exchequer or rate fund contribution; the total rent requirement is then spread over all the properties of the authority. Individual house rents thus reflect the average historical cost of an authority's entire housing stock.

The cost of new council housing is heavily disguised under the present financial scheme. Since rents are pooled and based on outdated housing costs, the actual cost of new capital programmes can be subsumed in the housing revenue account and deferred through long-term borrowing arrangements — frequently 60 years. At the same time, the Exchequer grant system in no way guarantees help to needy tenants, for major subsidies are received by the local authority and used to subsidise houses rather than tenants. Moreover, since allocation methods are determined locally, with eligibility and points systems differing greatly even between adjacent housing authorities, there is no obvious relationship between existing houses, new houses and the needy. This is because the financial circumstances of families are not explicitly included in local authority criteria of provision, which are based on the prospective tenant's family circumstances and existing housing conditions. Tenants, once allocated a house, are normally permitted to continue as local authority tenants for life, even though their needs may alter radically.

Exchequer subsidies

In 1980–1 central government subsidies to local housing authorities totalled £2,029m, but as shown in Table 4.3, had fallen in 1983-4 to £656m. This is a considerable decline and the real rate, after inflation adjustment, is even greater. The general rate fund net contributions, although high, have risen only marginally in the 1980s, from £523m in 1980–1 to £636m in 1983–4. In consequence, the general rate funds have not bridged the gap arising from the decline in Exchequer subsidies. Instead, to balance the HRAs nominal rents have increased at a rate faster than the retail price index. These higher rents, in turn, have meant higher payments to people under the housing benefit scheme in line with the government's declared wish to switch from subsidising housing *per se*, to helping people pay for housing. The effect has been that only 50 out of over 300 authorities in England and Wales were in receipt of Exchequer subsidy by the mid-1980s. As discussed earlier, local authorities must now balance their accounts through either rent charges or rate fund contributions and, so long as they operate within the financial framework laid down by government, authorities now have greater

Table 4.3: Local and central government support for housing in the UK, 1979–80 and 1983–4

£ million

	1979–80	1983–4
Assistance to supplementary benefit recipients:[a]		
Rent	500	1,700
Mortgage interest	50	150
Rent rebates and allowances[b]	300	900
Improvement and thermal insulation grants[c]	189	1,266
Central government subsidy to LA and new town housing[d]	1,828	656
Rate fund contribution to LA housing	404	636
	3,271	5,308

Notes: a. 1979–80, based on amount of rent and interest taken into account in assessing supplementary benefit entitlement. 1983–4 certificated rent rebates and allowances and mortgage interest; including housing benefit supplement; b. Rebates and allowances to non-recipients of supplementary benefit; c. Grants paid to private owners and tenants; d. Including subsidies to Scottish Special Housing Association and Northern Ireland Housing Executive. (The table does not include the cost of mortgage interest tax relief for owner-occupiers, estimated at £4,750 million in 1985–6.)
Source: *Hansard*, 20 March 1985

freedom to exercise decisions on rents, management, and repair and maintenance costs.

Determination of rents

Council tenants, among others, are not only eligible to receive housing benefit, but local authority rents are by law required to be 'reasonable' both to the authority and to the tenants. The law is rather less precise in the legal interpretation and rent levels reflect political decisions of what is 'reasonable'. The essential consequence is that rent levels are not only below economic levels, but show a considerable variation from one housing authority to another. Since 1980, government has sought to encourage a rise in rent levels to reflect 'past and expected movements in incomes, costs and prices'. The result has been a 130 per cent increase in local authority rents between 1979 and 1984. However, public-sector

Table 4.4: Gross unrebated rents, 1978–86 (average amount in per dwelling per week)

£

	1978-9	1981-2	1982-3	1983-4	1984-5	1985-6
Local authority rents Unrebated rent charged	5.90	11.51	13.58	14.03	14.83	15.66
Registered fair rents[a] Housing association tenants	10.08	13.98	15.63	17.19	18.69	19.35
Private (unfurnished) tenants	8.38	12.40	14.11	14.85	16.71	16.85
Local authority rents as percentage of average weekly earnings[b]	6.6	8.2	8.8	8.4	8.3	8.1

Notes: a. These are the averages of rents registered during the corresponding calendar years; because of phasing of rent increases, the rents actually payable during those periods would have been lower; b. Average weekly earnings for full-time adult males, all occupations, GB.

Source: *The Government's expenditure plans 1986-7 to 1988-9*, Vol. 2, Cmnd. 9702, HM Treasury, HMSO, January 1986

tenants still spend a smaller proportion of their income on housing than any other tenure group. In 1983, average local authority rents were 5.1 per cent below those in the private unfurnished sector, 18.3 per cent lower than for housing association tenancies and 39.1 per cent lower than for regulated furnished tenancies.

The average unrebated rent for local authority tenants in recent years and also a comparison with average earnings and rents in other tenures is shown in Table 4.4. Following the increase between 1978–9 and 1982–3 it shows that since then rents have moved more in line with inflation. In 1986–7 it is assumed by government for subsidy purposes that rents will again rise broadly in line with inflation, by an average of 65p per dwelling per week. In future, the government has stated that rents are only expected to rise in line with inflation. Although the government attempts to adjust Exchequer subsidies among local authorities so as to encourage rent increases of 65p per week, in reality neither Exchequer subsidies nor rents reflect needs or costs of housing. This is because rent levels of a particular authority are more affected by history than by other forces. An authority with a large stock of houses built many years ago and hence with low outstanding debts, will have lower average costs and rent levels than an authority with a large proportion of new building. Varying maintenance costs and decisions about improvement, *etc* will also affect rent levels.

Historically, local authorities have been free to decide if and to what extent their housing revenue account should be subsidised from the rate fund account. Housing rents have therefore not been tied to the actual cost of housing but to their age, size, services and condition, *etc*. This has meant that differential rents between houses of different ages are not nearly as wide as would have been the case had rents been related to costs; and, conversely, it has meant that rents on new houses have been rather less than should have been the case. However, the degree of difference has been dependent upon the extent to which an authority has averaged the rents over an individual estate rather than over their total housing stock. This rental policy has meant that, at best, the housing revenue account has broken even, but of course few funds have ever been generated for new building, as would have been the case had market or replacement cost rents formed the basis of rent computation.

Pooled rents have also had implications for policy decisions. A local authority with rents based on aggregated outdated house values and costs could easily be biased towards new building because it would not have to charge a rent based on the cost of new houses or flats but on total historical housing costs including the new dwellings. Moreover, the low historical cost of funds also reduces the rents it would be necessary to charge on renovations.

The pooling system has also had other serious consequences. First, there has been no guarantee that Exchequer subsidies to meet housing costs have been going to needy tenants. Moreover, the individual circumstances of families have been almost ignored, leading to injustices between tenants of different family and financial circumstances. Second, subsidies, until recently, have been paid to local authorities and vested in buildings rather than to the individuals who might live there.

In sum, the adoption of rent pooling has serious implications for the efficient and equitable allocation of homes. The arbitrary allocation of houses decided on unique and locally determined non-price methods has not only paralysed any tenant sovereignty over the supplier but implied that efficient resource allocation has been undermined and superseded by political considerations. In short, there is neither consistency in the setting of rent levels between authorities nor rationale. It means that rents do not reflect differences either in the costs or popularity of different properties, and that similar properties in different authorities which might be expected to have similar capital values and letting values are let at very different rents. Of course, market-clearing rents would probably differ widely between authorities, but there would be an economic, and not political or historic, function for the differences.

ALLOCATION OF LOCAL AUTHORITY HOUSING

Among the housing authorities, allocation schemes adopted and practised vary widely, as each prescribes its unique definition of housing need as a substitute for the price mechanism and ability to pay. Indeed, little attention is paid to tenants' financial circumstances and it is only coincidental, although probably likely, that the tenant household may be poor. This is because available housing is probably not of high quality so that the better off don't want to live in it, and area amenities may be poor. Also people probably have to queue for the housing and the better off prefer not to wait. Length of residence, household size, age and composition, and housing conditions are the criteria by which tenants attract housing points and ultimate housing by the local authority.

The supply of housing is generated by new building, and by vacancies arising through the death or migration of existing tenants. Together these provide a finite supply of available dwellings to be rationed among prospective tenants on a waiting list. In addition, local authorities also have limited nomination rights to dwelling units provided by housing associations building within an authority.

Generally, local authorities are concerned with providing permanent homes for prospective tenants, although accommodation is maintained to provide temporary accommodation to those rendered homeless within an authority as a result of private landlord eviction, family separation, *etc*. Moreover, city and district authorities have a legal obligation to house homeless families on demand.

Over the period of 1979–80 to 1984–7, new lettings to local authority tenants in England and Wales remained fairly stable with about 450,000 lettings in 1984 and 1985; this included fewer lettings of newly acquired or modernised stock but an increasing number of

Table 4.5: Allocation of local authority housing in England and Wales, 1971–85

	1971	1976	1979–80	1980–1	1981–2	1982–3	1983–4	1984–5
New tenants (%)								
Displaced through slum clearance, etc	—	12	8[b]	8	5	4	4	3
Homeless	—	9	15	15	16	17	17	19
Key workers	—	2	2	2	9	8	9	8
Other priorities	—	7	8	6				
Ordinary waiting list	—	70	67	69	67	68	66	66
On non-secure tenancies[a]	—	.	.	..	3	3	4	4
Total	*100*	*100*	*100*	*100*	*100*	*100*	*100*	*100*
Lettings ('000s)								
To new tenants	257	287	288	291	265	271	260	255
To new tenants transferring:								
Within local authority	173	189	172	178	173	154	157	199
Other exchanges						38	35	
Total	430	476	460	469	438	463	452	454
Of which:								
New, acquired, or modernised stock	107	131	93[c]	85[c]	56[c]	41[c]	43[c]	44
Relets	323	346	367	384	382	422	409	410

Notes: a. As defined in Schedule 3, Housing Act 1980; b. Relets enquiry discontinued after 1977; data from 1979–80 are based on the housing investment programme returns, and therefore are not strictly comparable with earlier years; c. The number becoming available for letting within the year.
Source: *Relets enquiry – local authority housing, and housing investment programme returns*, Department of the Environment; Welsh Office

relets, as shown in Table 4.5. Most new tenants of local authority housing continue to be households on ordinary waiting lists; the proportion was 66 per cent in 1984–5 and had changed very little since 1978–9. However, over recent years there has been a gradual increase in the proportion of homes let to homeless households (17 per cent in 1984–5), and a decrease in the proportion of lettings to people displaced by slum clearance, while in addition four per cent of new tenants in 1984–5 were in non-secure tenancies.

In 1986, about 400,000 local authority dwellings in England and Wales (seven per cent of local authority stock) were classified as difficult to let. Over 80 per cent of such dwellings were in metropolitan areas; however nearly half of all local authorities state that no dwellings are difficult to let. It is likely that authorities define the term in varying ways.

Rationing mechanisms of local authorities are essentially twofold and each authority is empowered to vary its own terms and conditions, subject to statutory controls under the Homeless Persons Act. The first restriction on access arises from a prospective tenant's length of residence within an authority. There are wide variations in eligibility for waiting list registration. Some local authorities require that a prospective tenant either works or lives in the authority, while others require a minimum residence of five years in the authority. These eligibility rules have a major impact as a deterrent to tenant application in certain authorities, and to household mobility, and place additional pressure on adjacent authorities which may have less restrictive eligibility rules. This arbitrariness could be condoned if the eligibility rules actively reflected the housing problems of an authority, but frequently they do not. Instead, the eligibility rules represent the worst type of discrimination between people and their housing problems by protecting existing tenants within an authority through security of tenure to the detriment of outsiders, and abetting the immobility of all tenants.

A second rationing device basically operates once an applicant is registered and puts the onus on the prospective tenant to show their degree of need by reference to the number of points that they attract. Each housing authority devises its own points scheme for applicants and the weighting attached to different factors varies enormously between authorities. Generally, however, the factors can be broken down to existing accommodation, social, medical and house accessibility. For example, accommodation factors refer to lack of adequate family bedroom space, and the sharing and deficiency of kitchens, bathrooms and WCs, *etc*. Social factors embrace age aspects of tenants, forced family separation due to overcrowding, and other miscellaneous factors such as under-occupation in the case of the elderly. Medical points can also be awarded, normally by a District Community Physician. Finally, points may be given in respect of present house accessibility for prospective tenants currently sharing some or all facilities with other families or living with children in flats on upper floors. In any event, it is the local authority which is responsible for deciding on the factors considered in allocation schemes and the relative weight attaching to each factor.

Effects of allocation schemes

Local authorities have tried in their provision of public-sector housing to be kind to too many people. By setting rents low and providing security of tenure to people whose original need for public-sector housing has diminished, they have created a situation with vast excess demand. To ration the resources available they have to use rationing schemes which mean that housing goes only to those whose need is desperate or who have waited a long time. There are two possible remedies for this. One, advocated by many on the left, is a large addition to the resources available. The other is that local authorities should charge higher rents and be more strict about turning out tenants who no longer really need public sector housing; this would give them more vacancies, which would allow them to shorten waiting lists and to allocate houses to people whose need is less severe than that of those selected at present.

This is not to suggest that the market place is a perfect allocator of resources. However, non-price allocation schemes are invariably advanced as superior to market means. If there is a limited stock of housing to allocate, then it is necessary to decide somehow between the claims of A who is overcrowded, B who has medical reasons for wanting housing, and C who is disturbed at being too remote from their family.

It is hardly surprising that allocation schemes are frequently criticised for their complexity and apparent inflexibility, and foster cynicism and despair in people who believe that the waiting list really is a genuine reflection of society's concern for them rather than an interposed filtering mechanism.

Ongoing assessment of need of existing tenants is imperfect. Applications by tenants following the birth of additional children may imply a transfer to a larger house. But cases of under-occupation are rarely announced by tenants and are poorly monitored by housing officers. While it might be considered that the needs of would be tenants should be compared with those of existing tenants, the exercise is rarely undertaken in the light of the difficulty in forcing tenants out of accommodation even when their needs change radically. While local authorities enjoy greater legal power of eviction than private landlords, few authorities will evict other than for non-payment of rent. In any case, if they render a tenant 'homeless' they may be bound to rehouse. Thus, local authorities remain effectively responsible for the housing of families for life, housing perhaps originally on points reflecting household size and, years later, transferring the sole surviving occupier of the family to sheltered accommodation, irrespective of the changing needs of that family during a generation. The problems associated with compulsory eviction of tenants whose income and family circumstances may well no longer justify public housing has inevitably led some authorities to devise ways of encouraging vacancies. Such means include shared equity housing schemes and placing a greater burden of housing costs, given the low average re-lets, upon tenants through the sale of council houses. In neither case, however, do they enable local authorities to house those in real need for they simply shuffle the distribution of tenants and houses at best, and effectively admit the loss of a house from the local authority housing pool at worst.

DRAWBACKS WITH COUNCIL HOUSING

Post-war policies in respect of council house building and allocation have had a number of important effects.

Excess demand

In spite of a massive long-term programme of building, excess demand for council houses remains. Excess demand, represented by waiting lists, not only ensures a sustained pressure for greater house-building, but places local authorities in a political and financial straitjacket from which they cannot escape to devise an effective and probably lower-cost programme for the equitable and efficient allocation of housing. Indeed some local authorities have been accused of deliberately trying to inflate their waiting lists so as to provide the rationale for building more houses. At the same time, local authorities as landlords have enjoyed almost complete producer sovereignty, and tenants have been denied the opportunity to assert any power in the market for public rented housing while it remains subject to arbitrary and local political decision. It has only been in the 1980s that right-to-buy legislation has offered effective choice to some existing tenants.

Quality

There is considerable evidence of dissatisfaction amongst tenants with council housing provision and standards. Design features have been poor or inadequate and tenants dislike certain estates and so abuse facilities. These deficiencies, it is argued, lead to vandalism and crime. That much public housing has been inappropriately or badly designed and built is no doubt true. While the faceless planners, architects and housing managers of the 1950s and 1960s may bear some of the criticism, little attention is paid to the inevitable consequences of local authority housing hegemony as it has been practised over many years. The politicisation of housing has been concentrated on the vast council estates where identical, drab and featureless buildings have contributed to the extension of the urban sprawl. Moreover, design has not been the only short-coming of council house building. There are serious structural defects in many council houses, including those built with prefabricated reinforced concrete (PRC) components before 1960. With some 140,000 needing drastic repair, the total cost is estimated at some £1.5b. The repair or replacement of these houses could account for a considerable share of new building finance.

Mobility

The effects of allocation policies on tenant mobility and choice are difficult to test but can be logically deduced. Mobility measurements of all tenures show that local authority tenants move least. In view of the problems facing any local authority tenant it is surprising that they are able to move at all. Inter-authority moves are notoriously difficult to effect. Unless local authority tenants can arrange a mutual exchange on a quasi-barter basis, their only recourse is to

61

move to private accommodation in the authority of their choice and apply for registration. Only if they know that the points accumulated are at least equal to the current letting levels of that authority can they be reasonably assured of a house at some future date, depending on the residence qualification to be fulfilled. Tenants may attempt to render themselves legally 'homeless'. While some authorities do provide housing bureaux or advice centres, many do not and in the absence of nationwide official bureaux, tenants revert to newspaper advertising, *etc* in the hope that perhaps bartering will provide an earlier solution than the ponderous mechanism of bureaucratic allocation and exchange.

Tenant choice of the area or type of council housing is highly restricted by the allocation policies of local authorities. Applicants who refuse to accept a house, perhaps because it is in a poor area, do so at their peril and are usually wiser to accept and then apply for a transfer in the hope that the bureaucratic machine will eventually grant them a home in a desired location. In any event, however, the tenant's subsidy, being based on the reasonable rent of the house, is not totally transferable, whether within an authority or to another. Tenants who move out of local authority housing altogether lose that subsidy. Given the effects of present allocation policies on immobility, one suggestion for reform might be that tenants who do wish to move locations and tenures could well be better served if local authorities were empowered to provide cash grants to tenants, equivalent to their present council house subsidy, as an increased financial inducement to permit mobility and increase local authority re-lets. However, such a cash grant, measured by the difference between accounting cost and rent, may not be sufficient to promote mobility in cases where the market price of housing is greater than either.

NOTES

1. Capital receipts are paid into the housing revenue account and invested, with interest earned paid into the account.

5

Private Rented Housing: Rent Control in Theory and Practice

DECLINE AND FALL

Private rented housing in the UK provides a relatively small share of the total stock of housing and contains a high proportion of pre-1919 housing. It consists largely of flats and terraces with low space and amenity standards and is heavily concentrated in inner city areas. Rented accommodation has the merit for tenants that no capital is needed and there is no responsibility for major repairs or, depending upon the conditions of the lease, internal or external maintenance. Tenants pay for the housing services consumed and do not bear the considerable responsibilities associated with house ownership. On the other hand, tenants may experience a lack of control over the quality of accommodation available. During periods of inflation, tenants do not own an inflation-proof asset, and may well be subject to considerable increases in rent. Whilst there are merits and demerits to renting, the demand and supply of rented accommodation has been, for many years, heavily influenced by the imposition and effects of rent control.

Rent restriction

First introduced as an emergency measure to prevent profiteering by landlords during the First World War in areas where housing shortages (and rising rents) emerged as a result of labour inflows for munitions production, controls have remained ever since. A succession of Rent Restriction Acts during the next 40 years reflected the then political view that rent control was a necessary, albeit temporary, imposition which could be repealed at the appropriate moment. The Rent Act of 1957 provided the first element of decontrol leading to rent increases for some five million controlled tenancies and block decontrol of properties with high rateable values and those properties falling vacant which were relet. However, this Act and its provisions were shortlived; many of the original intentions were never actually implemented and no further substantial measures of decontrol have so far been enacted.

The UK private rented sector is practically unique in having so much complex legislation formulated and implemented by successive governments on the basis of political dogma and unsubstantiated conventional wisdom. Ironically, the increase in size and complexity of the various rent acts has been inversely related to the size of the privately rented housing sector.

The Rent Act 1965, which was later consolidated with other existing legislation in the Rent Act 1968, was a major influence on the supply of rented accommodation between 1965 and 1974, for it made vital the distinction between unfurnished and furnished housing. All lettings by private landlords of unfurnished dwellings with a rateable value at the time not exceeding £400 in Greater London and £200 elsewhere were to become regulated tenancies and subject to 'fair rent' registration and security of tenure. This security provided for a regulated tenancy to be transmitted twice to members of a tenant's family at death. 'Fair rents' were to be determined by rent officers and ratified by a rent assessment committee. The 1965 Act distinguished three separate markets for rented accommodation — the regulated unfurnished market, the unregulated furnished market, and the old controlled housing — which were to be gradually phased into a new regulated sector.

Of critical consequence for rental housing under the 1965 Act were the determining criteria laid down for the assessment of a 'fair rent'. These state that:

a) In determining for the purposes of this Act what rent is or would be a fair rent under a regulated tenancy of a dwelling house regard shall be had, subject to the following provisions of this section, to all circumstances (other than personal circumstances) and in particular to the age, character and locality of the dwelling house and to its state of repair.

b) For the purposes of the determination it shall be assumed that the number of persons seeking to become tenants of similar dwelling houses in the locality on the terms (other than those relating to rent) of the regulated tenancy is not substantially greater than the number of such dwelling houses in the locality which are available for letting on such terms.

Under (a), a 'fair' rent is determined by reference to the attributes of the property and not to the financial or personal circumstances of the tenant. Under (b), the scarcity of accommodation is seemingly to be ignored. The intention, however, would appear to be that only 'abnormal' scarcity should be ignored and the rent set at a level where all 'need' is met.

In the context of a perfectly competitive market the absence of such 'scarcity' would be achieved in the long run at the equilibrium level. The implication of such an interpretation is that, unless long-run equilibrium (assuming a perfectly competitive market) is already achieved, the 'fair' rent is permanently less than the market-clearing one and conditions are created such that there is no incentive for landlords to increase the supply of accommodation. Moreover, the application of 'fair' rents and the introduction of rent allowances in 1972 and later housing benefit may be interpreted to mean that a 'fair' rent is a rent that families can afford to pay.

Under the Rent Act 1974, there was an extension of the legislation already governing unfurnished accommodation, to the furnished sector. On the request of either landlord or tenant 'fair rents' would be assessed by rent officers and security of tenure automatically extended. Although the Act was originally conceived as a way of controlling a hitherto relatively buoyant free market, the effect of the legislation was to create new 'loopholes' (as the Rent Act of 1965 had before it). No longer was the legislative distinction between furnished and unfurnished accommodation a determinant for regulatory purposes; it now rested upon the residential status of the landlord. Since 1974, a resident landlord, subject to certain conditions, has been able to claim exemption for this property, but almost all other accommodation is subject to some rent regulation and security of tenure provisions. The only exemptions are in the cases of lettings by educational institutions and accommodation for bed and breakfast and holiday letting purposes.

In the latter half of the 1970s, the then Labour government had become concerned at the effect of the Rent Acts of 1965 and 1974. In the *Review of the Rent Acts (1977)*, the Labour government concluded:

There is a good deal of criticism that the rent acts inhibit the existing stock of houses from being used to full advantage or maintained in a proper condition. It is contended that the effects of the acts in practice are sometimes unfair to tenants or landlords, or exclude prospective tenants and discourage landlords from keeping properties available for letting. There can be no doubt that the complexity and obscurity of the acts are a source of frustration and anxiety to landlords, tenants and those responsible for administering and interpreting the legislation.

In that Review a number of objectives that the then government believed should be met in any new proposals to halt the decline in the number and quality of rented houses were specified:

'(a) To safeguard the interests of existing private tenants;
(b) To ensure that fit private rented houses are properly maintained and kept in repair;
(c) To promote the efficient use of housing and to encourage, for example, the letting of property which might be available for only short lets;
(d) To ensure that the methods and criteria for the determination of rents are tailored to meet the difficulties faced by landlord and tenant;
(e) To simplify the law on private renting and to make for a speedier and more effective resolution of landlord/tenant disputes;
(f) To provide for a legislative framework which maintains a fair balance between the interests of tenants and landlord so that private rented accommodation can contribute effectively to meeting housing needs and choices whilst evolving into social forms involving, and acceptable to, existing landlords and their tenants.'

Quite how such diverse and mutually exclusive policy objectives could be achieved in any reform of legislation was never articulated, yet alone implemented. Two years later, in 1979, the Labour government was succeeded by a Conservative government. In view of the Conservatives' past objections and hostility to rent control and a manifesto which had expressed a strong predilection for market forces, any expectation that rent control would survive into the 1980s seemed improbable. In fact, it has survived to date (not entirely intact perhaps) and the vast majority of rented property remains under statutory control. By the mid-1980s, rented housing had the following features:

a) It has undergone a sustained decline in the number of dwellings, both absolutely and relatively, as gradual replacement and augmentation by local authority and housing association housing and owner-occupation has occurred. In 1914 about 90 per cent of the total housing stock was privately rented, but by 1939 this proportion had fallen to 58 per cent. This relative decline may be attributable mainly to the rapid increase in owner-occupation and local authority housing, for the absolute net decrease in privately rented houses was only of the order of 0.5 million in the inter-war period. While 1.5 million houses were sold for owner-occupation, nearly a million new or converted houses were added to the private rented sector during the inter-war period. Since the Second World War, sales for owner-occupation have been important, but in contrast to the inter-war period, there have been only 0.4 million new or converted houses added to the private rented sector. By 1976 the private rented stock, which had stood at 7.1 million in 1914, had fallen to 2.8 million, representing only 15 per cent of total stock and providing housing for some 20 per cent of all households. During the next ten years private rented housing has continued to decline to a mere ten per cent of dwellings at the end of 1986. During the 1980s, the number of private tenancies has been decreasing at the rate of 100,000 homes a year.

b) A wide diversity of households in rented housing is revealed in successive censuses. There is a considerable contrast in characteristics of tenants of unfurnished and furnished accommodation. The unfurnished sector comprises mainly the elderly, low-income households and tenants who have occupied their homes over many years. By contrast, the furnished sector comprises mainly the young, the single, above-average income and fairly mobile households. Similarly, landlords of unfurnished property tend to be in the older age groups and have generally either inherited or purchased their let property many years ago. Their low incomes (often pensions) and the low-income yield on their investment (notwithstanding availability of improvement grants), have precluded substantial improvement and maintenance of their properties. In the case of furnished tenancies, a large number of landlords are either resident or company landlords. The balance comprises landlords who let on a largely *ad hoc* basis, owning one or two properties.

c) The geographical distribution of the private rented sector is uneven and tends to be concentrated within inner city areas. London alone accounts for over 25 per cent of private rented dwellings and about 37 per cent of all households in furnished tenancies. A number of the larger university cities also contain a relatively high proportion of private lettings.

d) According to the 1981 House Condition Survey, 63 per cent of private rented dwellings were built before 1919 and the majority of the balance were built in the inter-war period. This is a major reason for the disproportionately high and widespread disrepair of rented housing. Privately rented dwellings now account for over 30 per cent of all unfit homes and over 30 per cent of those lacking one or more basic amenities. Private sector tenants are 16 times more likely to live in a home which is in major disrepair than those tenants living in the public sector. Finally, furnished tenants also pay out a higher proportion of their income for housing purposes than any other group.

e) The rate of return is extremely low and the decline in rental housing has been prompted by a diminishing, and sometimes negative, rate of profit on rented property.

In 1985, 'fair' rents varied from an average £29.88 a week for furnished tenancies in London to £15.55 for unfurnished tenancies in Scotland. They provide an average gross return of only three to four per cent on the vacant possession value of property (in London this return is as low as one per cent; in Cleveland as high as seven per cent). Most landlords would get a better return if they were to sell up and invest the money in a building society.

However, 'legal' loopholes do exist. For example, an increasing number of landlords avoid rent controls and security of tenure provisions by letting properties on short-term 'licences'. A recent Greater London Council survey found that over half of all new lettings in London were now outside rent act controls.

The decline and fall of the UK private rented housing has frequently been blamed on imposition and perpetuation over 70 years of rent control. What are the theoretical effects of rent control and the evidence to support this contention?

THEORETICAL MODELS OF RENT CONTROL

In an analysis of the general consequences of rent control two main models with restrictive assumptions may be advanced. The first model employs a neoclassical but *ad hoc* conventional supply and demand analysis of the effects of control in the short and long terms under the assumption of homogeneous units and sites with a control system that imposes a ceiling on the rent of the dwelling units. The second model of the rented housing market assumes that homogeneous housing services, rather than homogeneous housing units, are traded in the market and also specifically distinguishes between the short and the long run.

These two models elucidate many of the aspects of rental market controls. They are simple and 'partial-equilibrium' in character, and capture the fundamental elements of the rental housing market and how it will react to controls on rents and evictions. Their predictions often conflict with those drawn from more informal analysis or anecdotal commentaries of the effects of rental market controls. It is very important to have a formal framework or 'model' as a background to a discussion of intervention in any market.

Explicit model with quantity variable

In a simple neoclassical equilibrium housing model, Figure 5.1, of the short and long run effects of rent control the following assumptions are made:

Figure 5.1: The market for housing

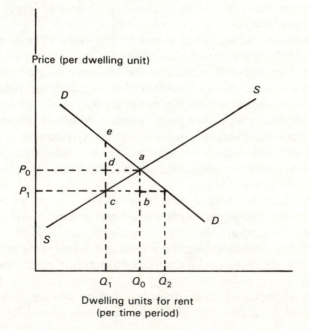

(a) Dwelling units are homogeneous and have the same site values: that is the level of housing services yielded by each dwelling unit is not variable;

(b) The rental market for dwelling units is perfectly competitive;

(c) *DD* is the market demand curve for dwelling units and slopes downwards;

(d) *SS* is the market supply for dwelling units and slopes upwards on the presumption of increasing costs. Since the market model is not derived from an explicit model of the firm, no specific reference is made to whether *SS* represents short or long run. More realistically, *SS* may be presumed to be perfectly inelastic in the short run, on the grounds that the number of dwellings units cannot be altered due to (i) the smallness of annual production in relation to total market stock, (ii) the spatial immobility of houses, and (iii) the high capital/output ratio of housing.

It is assumed in Figure 5.1, that the market is in equilibrium with Q_0 dwelling units and a rent of P_0 each. A ceiling rent of P_1 which is below the equilibrium rent of P_0 for all dwelling units is then applied. It could be considered more realistic to postulate the freezing of rents at P_0, and then to examine the effects of an increase in demand on the housing market — for example, increased household formation. Such an approach is adopted by Lindbeck (1967) but does involve explicit assumptions on the price elasticity of supply in the short and long run.

Following the imposition of rent P_1, the model yields a number of important predictions for the short run:

(a) The number of occupied dwelling units will decline to Q_1;

(b) There will be excess demand of $Q_2 - Q_1$ dwelling units. Such excess demand can conceivably lead to any one, or all, of the following consequences: (i) non-price rationing (tenant discrimination), (ii) increased owner-occupation, (iii) the development of black markets, (iv) increased demand for housing in other rental sectors (local authority housing), (v) multiple accommodation, and (vi) increased homelessness;

(c) There will be a transfer from landlords to tenants of area $P_0 dc P_1$ and deadweight losses of economic surplus of *ead* (loss in tenants' surplus) and *dac* (loss in landlords' surplus).

(d) Any subsequent increase in demand will lead to increased excess demand, but would not affect the number or rent of dwelling units;

(e) Any increase in landlord costs will shift the supply curve upwards and lead to a further reduction in dwelling units offered for rent.

The effects in the long run will be more pronounced than in the short run. The long-run supply curve will tend to be more elastic with respect to rents, *i.e.* it will be 'flatter'. Under the assumptions of this model, the long-run supply curve will be horizontal or 'perfectly elastic'. This is because landlords have more options open to them in the long run. Under rent control more of them will remove their properties from private renting. There will be fewer dwellings let, greater excess demand, *etc.* In this simple model, the setting of a rent ceiling below the minimum long-run average cost of the least-cost landlord may well result in the total demise of the private rental market.

Landlords' incentive to exit the industry, perhaps completely, is normally anticipated by legislators, who usually combine rent controls with restrictions on evictions. In Figure 5.1, what will be the effect if all initial Q_0 tenancies are now protected by security-of-tenure provisions in the legislation? Actual quantity of dwelling units does not change, but the lower rent of P_1 induces demand of Q_2 so that there is still an excess demand but of only $Q_2 - Q_0$. The consequences are less pronounced than under rent control alone, although landlords suffer more. The transfer from landlords to tenants is now P_0abP_1, compared with P_0dcP_1 under rent control alone. The deadweight losses of landlords' and tenants' surplus under rent control, *dac* and *ead*, respectively, disappear under the rigid form of rent and eviction control. On the other hand, if there was an increase in demand for housing, this would immediately result in a return of the deadweight losses and an increase in excess demand as there will be no supply response at all. In effect, society will be deprived of the production of units of housing that are worth more to that society than what they cost to produce.

Explicit model with quality variable

Originally developed by Frankena (1975), this model achieves a greater reality for policy prescriptions in the face of rent control. It not only incorporates an explicit model of the individual firm which provides housing services, but distinguishes between the short and long adjustment period and so allows quality to be a variable. The specific assumptions are as follows:

(a) Homogeneous housing services are traded in the perfectly competitive rental market, rather than the services of homogeneous housing units as in the previous model (see Olsen, 1969a);

71

(b) The long-run market supply curve, S_1S_1, is perfectly elastic since all firms have the same minimum long-run average cost curve (*LRAC*);

(c) The short-run market supply curve, S_2S_2, is constructed by summing the short-run firm supply curves in Figure 5.2;

(d) The short run is defined as the period where no new firms can enter or exit the market and existing firms can only adjust quality, but some costs are fixed. Figure 5.2 incorporates the short-run average cost curve (*SRAC*), the average variable cost (*AVC*), and the short-run marginal cost (*SRMC*). These have the shapes conventionally assumed in microeconomic analysis.

In Figure 5.2(a), it is assumed that initially the firm is in equilibrium producing q_0 housing and operating at zero profits (*i.e.* 'normal' profits only) and that the market for housing services is in long run equilibrium, so that Q_0 units of housing services are traded as a rent of P_0 per unit.

As a result of a maximum rent of P_1 price per unit of housing services, the model yields the following predictions:

(a) a decline in housing services in that the representatives profit-maximising firm will allow its dwelling units to deteriorate until the flow of housing services declines from q_0 to q_1 in Figure 5.2, and the aggregate output of housing services will decline from Q_0 to Q_1 in Figure 5.2(b). Any resulting lack of maintenance and repair is worsened by the landlord's uncertainty as to whether he will be able to repossess the property, given tenants' security of tenure.

(b) Rents are depressed below market levels. Since the price per unit of housing service will decline from P_0 to P_1, the rent per dwelling unit will also decline — this is because both the number of units of housing service per dwelling unit and the price per unit of housing service will decline.

(c) Excess demand for accommodation is represented by Q_2-Q_1 units of housing service. Rent control poses a problem for potential new entrants to the market and for existing tenants who want to move. In such a restricted market, the cost of searching for accommodation can be very considerable, with complex effects on the welfare of the potential tenant.

(d) Any rise in costs will depress housing services. Any increase in the demand for housing services will increase excess demand, and any increase in variable costs after the imposition of rent control will cause a further decline in housing services.

(e) Landlords will subsidise tenants. To the extent that tenants pay low rents below the market price they are in effect receiving a

Figure 5.2: Equilibrium in the housing market

(a)

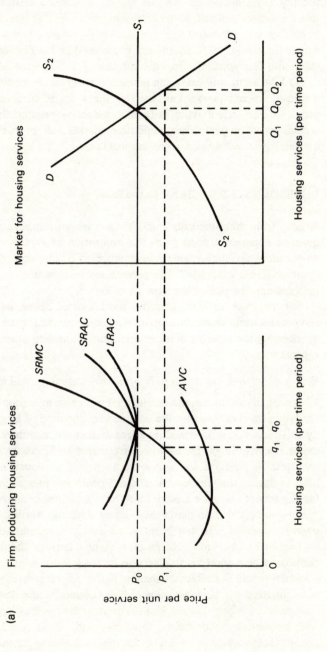

subsidy from landlords. This is clearly a random distribution of income from landlord to tenants without regard to the financial circumstances of either. If the state really believes that tenants should be subsidised, it should ask if it should be the landlord or the state who is responsible for this subsidy.

(f) Landlords will leave the market. Unless there are barriers to obtaining vacant possession, such as the strict security of tenure provision currently existing in Britain, the model predicts that under conditions of a perfectly competitive market, all properties will eventually be withdrawn from the market.

CONSEQUENCES OF RENT CONTROL

Arising from these theoretical models a number of important consequences apparently arise from the imposition of rent control. To what extent are they realised in practice in Britain and elsewhere from available evidence? The predictions of rent control and the implications for policy are now assessed.

For as long as rent controls have existed there have been economists critical of their introduction, perpetuation and consequences for the housing stock and its quality, among other adverse aspects.

Rent control reduces the supply of accommodation available

Rent control diminishes the incentives for tenants to 'ration' accommodation for it encourages the continued occupation of large houses by small families who would, in a free market, either sublet or move to smaller dwellings. Faced with a fixed rent and inadequate return on capital, a landlord will adjust the quality of the housing services offered during the tenancy. A landlord could sell to a sitting tenant but if a tenant leaves, a landlord will have a greater incentive to sell with vacant possession than to seek a new tenancy. Abolition of rent control would at least put property on an equal footing with other investments and there is no reason why new lettings should not be stimulated at market rent. The assertions of those who argue that abolition would not affect the decline in the private rented sector are not supported by the experience of those countries that do not have rent control. Where the stock of accommodation is small compared with demand by households, rents rise and people are forced to economise on space, *e.g.* house sharing, etc. Moreover, high rents stimulate the conversion of large houses into flats and also new construction. Those at lower income levels also benefit by the increased supply of accommodation in being able to filter upwards into better accommodation as it becomes available.

For example, Friedman and Stigler (1946) in their important analysis of the effects of rent control in the USA contrasted the effects of rationing by price with rationing by choice and favouritism. Under a price system '. . . if demand for anything increases, competition among buyers tends to raise the price. The rise in price causes buyers to use the article more sparingly, carefully and economically, and thereby reduces consumption to the supply. At the same time, the rise in price encourages producers to expand output. Similarly, if demand for any article decreases, the price tends to fall, expanding consumption to the supply and discouraging output.'

This mechanism is illustrated by Friedman and Stigler (1946) when, following an earthquake, San Francisco used the free market method to deal with its housing problems, with a consequent rise of rents. They concluded that:

(a) In a free market, there is always some housing immediately available for rent — at all rent levels;
(b) The bidding up of rents forces some people to economise on space. Until there is sufficient new construction, this doubling up is the only solution;
(c) The high rents act as a strong stimulus to new construction;
(d) No complex, expensive and expansive machinery is necessary. The rationing is conducted quietly, and impersonally through the price system.

Rationing by chance and favouritism arises from supply effects in cases where legal ceilings are imposed on rents. This will inevitably lead to a reduction in the number of places to rent. The main supply effect is a reduction in the efficiency with which housing is used by those who are not forced to double up in the face of shortage. Incentives to economise space are much weaker in conditions of control, because rents are now lower relatively to average money incomes. Moreover, the scarcity resulting from rent ceilings imposes new impediments to the efficient use of housing. A tenant is unlikely to abandon his 'overly large' home to seek accommodation of a more appropriate or desired size. Clearly this feeds back to the demand side in that every time a vacancy does occur the landlord is likely to give preference in renting to smaller families or the single. Friedman and Stigler argue that the removal of rent ceilings brings about a doubling up in a quite different manner. In a free rental market tenants would yield up space if they considered the sacrifice of space to be repaid by rent received. Doubling up would be by those who had space to spare and wanted extra income, not, as now, by those who act from the sense of family duty or obligation, regardless of space available or other circumstances. Within the supply effects of rent control, Friedman and Stigler conclude that ceilings cause haphazard and arbitrary allocation of space, inefficient use of space, retardation of new construction and indefinite continuance of rent ceilings, such as experienced in the UK, or subsidisation of new construction and a future dearth of rented accommodation.

Following considerable empirical study of the Swedish housing market, Rydenfelt (1980) reached a similar conclusion on the supply effects of control. He argued that the 'popular opinion' encouraged by defenders of rent control that the Swedish housing shortage was a product of the war does not accord with the evidence. All of the data indicate that the shortage during the war years was insignificant compared with the shortage after the war. It was only in the post-war rent control era that the housing shortage assumed such proportions that it became Sweden's most serious social problem.

Maclennan (1978) examined the effects of the Rent Act 1974 on the furnished rental housing sector in the city of Glasgow, and conducted a rigorous analysis of the behaviour of landlords and tenants before, after and during implementation of this act. He found that the letting behaviour of landlords can be interpreted as a desire to continue 'letting, but to reduce the risks of furnished renting'. According to Maclennan, such rational behaviour on the part of landlords meant that legal landlord behaviour could in the long run, frustrate one objective of the 1974 Rent Act — that was to ensure a supply of furnished private lets with full security of tenure for poor families within the sector.

Maclennan was unambivalent in his conclusion of the dire supply effects of the 1974 Rent Act. He argued that it did indeed reduce the supply of furnished property, but that this reduction was achieved by the ability of landlords to alter their product or tenantry and so escape the restrictions of the act. He went on to argue that if the 1974 Rent Act was intended to discourage private renting further, then it could be said to have been an unequivocal success. Moreover, he considered this view failed to consider the costs and benefits of rental sector decline. In particular, the reduction in private rental supply was not matched by an expansion in municipal property available to the typically young, transient tenants of the sector. Given the rising price of property which precludes entry into the owner-occupied sector, furnished sector tenants, at least in Glasgow, became trapped between the effects of the Rent Act and what Maclennan saw to be intransigent policies on municipal letting.

Rent control can be seen to result in a 'shortage' of rental accommodation as a consequence of both supply-side and demand-side effects. On the supply side there will be a reduction in the total amount of rented housing available and a decline in the efficiency of its use. Moreover, according to the empirical evidence, rent control also causes an expansion of demand. In some cases — San Francisco in the 1940s and post-war Sweden for example — Friedman and Stigler and Rydenfelt respectively suggest the ironic phenomenon of total growth of construction activity actually outpacing population growth while the measured housing shortage is increasing!

Rent control militates against the maintenance and improvement of housing quality and standards

Under conditions of rent control a deterioration in the quality of dwellings may be expected. If only rents are controlled then landlords will probably try and eject tenants and sell the property on the free market for owner-occupied dwellings. As a consequence of eviction controls this may not be possible so it may be in their interests to reduce or eliminate spending on repairs and maintenance both as a means of 'cutting one's losses' and as a stimulus for the tenant to quit the property. Incentives facing the landlord with regard to repairs also include expectations about the future of the controls on the market. If controls are expected to be very long-lasting (and sitting tenants expected to remain in occupancy) the landlord may even allow structural damage to go unattended. A number of empirical studies attest to these consequences *e.g.* Rydenfelt (1980) on Sweden, de Jouvenal (1948) on post-war France and Paish (1952) on Britain. Perhaps the most famous remark is that of Assar Lindbeck (1967) who said, 'In many cases rent control appears to be the most efficient technique presently known to destroy a city except for bombing.'

All these writers are highly critical of the effects of rent control on housing quality. Any rent level which denies a return competitive with alternative investments, encourages neither new construction of property to let, nor new lettings from the existing stock of houses.

If rents are held down during periods of rapid increase in the price of new building and the repair and maintenance of existing stock, the position is worsened. When rents lag behind price changes in this way, landlords have every incentive to reduce the 'housing service', *i.e.* the quality of housing they provide, particularly as far as physical aspects of property are concerned. As discussed earlier, fair rents currently yield an average gross return of less than three to four per cent in England and Wales, although the rates of return vary between even lower yields in London and somewhat higher yields in other parts of England and Wales. Net returns are correspondingly lower and evidently inadequate to keep existing landlords in the market, particularly if they have an opportunity for sale with vacant possession to the owner-occupied sector. Returns derived from private renting under regulated rents are not adequate to justify repairs or improvement to property. The effects are that the lack of maintenance and repair are aggravated by the landlord's uncertainty as to whether he will be able to repossess the property, in view of the provisions that currently exist granting tenants security of tenure.

Rent control constitutes an indiscriminate subsidy from landlords to tenants

Rent restrictions have obvious implications for income distribution. To the extent that tenants pay rents below the real market price, they are in effect receiving a subsidy from landlords. Without accurate data, it is impossible to measure the effects of this. What is clear is that a random redistribution of income from landlord to tenant takes place without regard to the financial circumstances of either. If a government really does want to subsidise tenants in this way, it should be the government itself, rather than the landlord, which is responsible for this subsidy.

In the absence of rent control, households of differing income levels would optimise their consumption of housing and other goods in relation to their income. Under controls, income redistribution between landlords and tenants takes place and there is no guarantee that the households which gain are more deserving than others.

Rent control accentuates the discriminatory effects of the fiscal system

For taxation purposes landlords are treated as businesses — but not as generously as ordinary business in the treatment of capital asset depreciation. It is assumed that houses last for ever and maintain their value. Thus no depreciation on buildings is allowed for tax purposes. Landlords must pay tax on rent income net of management expenses, including interest payments on any loan, and let properties are subject to capital gains tax. Both these latter provisions are in marked contrast to the treatment of owner-occupiers for tax purposes, *i.e.* no tax on imputed income from owner-occupying and no tax on capital gains.

Rent control hampers new tenants entering the rented sector

Rent control poses as big a problem for new entrants to the market, and for existing tenants who want to move. In such a restricted market, the cost of searching for accommodation can be very considerable, with complex effects on the welfare of the potential tenant, discussed below. The effects of conditions of permanent excess demand in the market can lead to serious unintended behaviour by landlords.

Rent control allows landlords to discriminate between tenants on non-price grounds

Under rent control and conditions of excess demand discrimination is rife, black markets exist, and winkling [1] is common. Allocation is arbitrary and inequitable. First, landlords faced with fixed fair rent and prospects of long-term security of tenure reduce risk by choosing tenants by methods other than price. As a risk-averter, the landlord faced with a fixed rent will probably discriminate, preferring tenants who are likely to be mobile rather than those with families. Abolition of rent control would aid many of those currently discriminated against to find rented homes through the payment of market rents. Second, excess demand is such that willingness-to-pay for the available quantity is greater than the controlled price, and this encourages 'black market' activities. Third, landlords often have an incentive to get rid of tenants even if they are 'good tenants' — the property is worth more without them than with. Rather unpleasant methods may be used to eject tenants. All of these repercussions of rental market controls have been observed in many instances where this type of legislation has been introduced. We now examine each in turn.

Discrimination can arise under conditions of non-price competition. In free-market conditions rent is the device which rations the available supply of housing. A landlord may require different rent levels to accept different types of tenants and, indeed, may set very high rents to deter those he does not want. The ability to offer a higher rent is a means of overcoming 'disabilities'. A potential tenant with a disadvantage can offer a higher rent in an attempt to secure a tenancy. If the landlord does not accept, it is either because the additional rent does not cover the anticipated additional costs of that tenant or he is willing to bear the cost of his prejudices. The more competitive is the market the more costly these prejudices can be. This is well illuminated by Friedman.

In Friedman's *Capitalism and freedom* (1962), it is noted that the businessman who expresses preferences in his business dealings is at a disadvantage compared to other individuals who do not. Such an individual is in effect imposing higher costs on himself than are other individuals who, in a free market, will tend to drive him out. In short, a free market is a device for limiting the degree of discrimination by bringing home its costs to the person practising it. In a price-controlled market, however, the situation is different. Because of excess demand as a consequence of rent control, the landlord will have numerous applicants for any available property coming onto the market. Even though potential tenants may be willing to pay an additional rent, there is, under rent control, no cost to the landlord in 'discriminating' according to his likes and dislikes. Minorities of all kinds — religious, sexual, ethnic, *etc.* — can be losers under these circumstances. Frequently, rental market controls include provisions making various types of discrimination illegal, but it is difficult to get evidence of discriminatory practices. Two important examples of non-price rationing may be cited. One is Friedman and Stigler's (1946) examination of San Francisco, where 'chance and favouritism' was the rationing procedure. The other example is from Canberra, where the Real Estate and Stock Institute of Australia (1975) has noted that, 'The rent controls will virtually eliminate accommodation for groups unless there is recognition by the authorities that a higher risk is involved for the landlord.'

The encouragement to abuse the law through black market dealings is a further major consequence of price ceilings. It is undesirable to bring the law into disrepute by encouraging law breaking — but this is precisely what price controls do. The imposition of price ceilings can create an interest for both the seller and the buyer to break the law.

If, for some reason, supply is fixed there will be a 'victim' from black market trading. The honest individual who does not break the law will tend to miss out in these circumstances. On the other hand 'sharp' operators will procure supplies at the expense of the weak, even if the latter are prepared to pay more than the controlled price.

Such illegality makes it difficult to get evidence. Occasional indications arise, such as Friedman and Stigler's (1946) claim that above-rent payments were made in San Francisco, and Cheung (1980) refers to the situation in Hong Kong during the 1920s, where 'the payment of "shoe money" had been a tradition in Hong Kong long before the ordinance was ever considered. Legend attributes the phrase to a polite euphemism for payments to middlemen who had to "wear out their shoes" in searching out living quarters for clients. By extension, the term can be applied to "courtesy" payments direct to landlords by prospective tenants trying to gain tenancy.'

Finally, landlords subject to severe restrictions on evictions may turn to other ways of removing tenants — some legal and others definitely not. These methods depend on the attractions of reletting at a market rent or selling in the uncontrolled market. Legislation is such that it provides considerable opportunities for the legal profession to find and exploit loopholes in the legislation on behalf of landlords.

Where legal efforts fail, landlords have resorted to unpleasant methods in order to eject tenants, such as threats of and actual violence, intimidation through late night 'anonymous' phone calls and other types of annoying behaviour. Numerous legal cases around the world attest to the extraordinary lengths to which landlords will go to repossess their property.

Rent control encourages avoidance by landlords

There are a limited number of exemptions for rent regulation legislation and these include holiday lettings, licences, lettings to companies, *etc.*. However, these exemptions have, not surprisingly, become massive loopholes exploited by landlords, to avoid rent control and security of tenure for tenants.

Given the present UK controls, households which cannot find dwellings to rent are forced to turn to 'substitutes' in the uncontrolled sector, *e.g.* lodgings, 'holiday' accommodation, *etc.* Interestingly, they are forced to pay a higher rent than if the whole market were completely free and also they have no security of tenure. In the absence of such a division in the market, total demand would equal total supply at the market-clearing rent. Some housing would always be immediately available, the quality enjoyed being determined by the rent people are able and willing to pay. Occupational mobility would be greatly improved.

Rent control reduces tenant mobility

Regulated rents evidently reduce the mobility of labour. If rents are fixed below their real market price, tenants will not only be reluctant to vacate existing regulated dwellings but will find it increasingly difficult to find comparable housing elsewhere. Not surprisingly, people will be extremely reluctant to move, even if they are unemployed and suitable jobs are available in other parts of the country. Other sorts of changes in circumstance that occur are things like the family growing up and leaving home; increases in the family size when children are born; the desire to take in a relative; the death of a family member; changes in income, *etc*. Over time, requirements are changing but the degree of adjustment to them under rent control is less than it would be otherwise. The situation becomes steadily worse as circumstances alter and as new tenants take what they can get rather than what they would prefer. It is only when the inconvenience is so great that the tenant will move voluntarily, in spite of the inability to find another controlled dwelling, that the house space is freed for another, perhaps more suited, tenant.

Even more than the inefficiency through the rigid use of house space is the cost to society caused by the disincentive for people in controlled dwellings to move in search of better job opportunities. This immobility has at least two adverse effects on efficiency. First, it can lead to greater mis-matching of workers to jobs as people forgo opportunities for better jobs, more suited to them. This is because of the likely existence of a stationary housing subsidy. Second, some workers will be unemployed and will remain idle in spite of job opportunities elsewhere. Evidently, the difference between their wage and the value they place on leisure activities is a loss to society as a whole. This lack of mobility has been noted often by experts who have studied the effects of rental market controls. For example Hayek (1929) argued that 'the restrictions on the mobility of manpower caused by rent controls mean not only that available accommodation is badly used to satisfy diverse housing requirements; they also have implications for the deployment and recruitment of labour, to which too little attention is paid.' Similarly, Paish (1952) in post-war Britain drew attention not only to the deleterious effects of rental market controls on labour mobility, but also to the effects on the efficiency of the use of house space, the importance of 'special favour' in procuring a tenancy and the existence of above-rent payments.

Rent control in Israel during the 1950s resulted in large disparities between legal and market rents. Schreiber and Tabriztchi (1976) drew attention to a practice developed in Israel in the 1950s where sitting tenants would introduce a new tenant to the landlord if they wished to leave. The new tenant would pay 'key money' which was shared between the departing tenant (two-thirds) and the landlord. This practice was initially illegal but became so widespread that it was legalised in 1958. It is perhaps worth noting that it does enhance the degree of mobility in a rent control situation and does provide some relief to the landlord.

Recently in Britain, the alleged existence of a 'north' and 'south' division of the nation has emerged as a potentially enormous problem. The decline of the old industries in the north of England and Scotland have been offset to considerable extent by the development of new industries in the south. However, while there is excess demand for labour in the south, this is not being met from the ranks of the unemployed in the north. The costs of labour immobility are, for an immobilised worker in employment, the difference between the worker's actual wage and that which could be earned in the best possible alternative occupation. The cost, for the unemployed immobilised worker, is the difference between the best wage possible and the value placed on leisure by the person. The aggregate cost of immobility would be the total of these differences over all persons concerned.

Housing policies have a major responsibility for the continuation of this imbalance. Minford, Ashton and Peel (1986) estimated the effects of all major housing market distortions — the Rent Act, subsidised council rents, tax deductibility of mortgage interest and environmental planning — on the overall level of unemployment. They find that if the rental market were decontrolled and council house rents raised to free-market levels, the level of unemployment would have fallen by about two percentage points. This is quite a significant result which would entail an increase in gross domestic product of the unskilled wage rate times the increase in employment. This could be about £3,500m (*i.e.* 500,000 × £7,000) in 1987 values. This figure does not represent the gain to society. It is an overestimate because it ignores the leisure foregone by the newly employed. However, it is an underestimate in that it omits the gain from a better matching of those presently employed with jobs. There would also be a significant gain from the more efficient use of the nation's stock of housing.

Conclusion

The effects of rent control are many and include reductions in the quantity and quality of the rental housing stock, the stimulus to various 'sharp', illegal and unpleasant practices; and the immobility effects on the movement of people. Albon and Stafford (1987) have argued that a succession of governments in various countries has been intent on establishing some of the effects of a collectivist economy in the microcosm. Apart from the manifest economic costs of rental market controls, they argue that it should not be forgotten that they involve a severe undermining of private property rights and an interference in the making of contracts. 'Under rental market controls there is a move towards collectivism. The rights to the income from and to make decisions about the use of private property are compulsorily reduced. The tenant develops an "interest" in the property without the landlord's consent. The property is, to some extent, removed from the landlord's control to the benefit of others. This is inconsistent with the basis of a free democratic society.'

Risks and other consequences associated with legislation evidently induce landlords to cease letting at the earliest opportunity, whilst those who continue to let may attempt to minimise the risk of financial loss by letting to the more transient groups of tenants — students, 'holiday-makers', *etc* and to reduce the standard of maintenance of their property. In any event, the justifications for rent control of successive governments have been compromised. Poor tenants are not protected, landlords are denied an economic return, properties are neglected and supply declines. Britain, unlike most other countries, still has a collectivist enclave in the rental housing 'market'. This has and continues to involve significant economic costs and much human suffering for very little benefit. These adverse effects of rent control provide an overwhelming case for early reform.

NOTES

1. Slang: to eject (as a winkle from its shell with a pin).

Part III

Housing Finance and Policy Reforms

6

Taxation and Owner-occupied Housing

MORTGAGE INTEREST TAX RELIEF

While political parties may have pondered the political desirability of retaining mortgage interest relief, no government, so far, has sought to remove it. Mortgage interest relief has been a major plank of governments wishing either to encourage, or at least be seen not to discourage, owner-occupation. Certainly, owner-occupation has become a highly desired tenure. While Britain does not have an exceptionally high percentage of owner-occupied housing, the average age at which the British buy their first homes is lower than that of people in almost every other industrialised country.

A recent survey (Building Societies Association, 1985) of international comparisons of housing tenure has shown that of households in Britain where the head was under 25, 30 per cent are owner-occupiers. Between the ages of 25 and 29, the comparable figure is 54 per cent. Households in Britain whose head is over the age of 45 are less likely to be owner-occupiers than similar households overseas. By contrast, in the 65 plus group, Britain has one of the lowest proportions.

This marked difference indicates the pace of the trend toward owner-occupation in the past 35 years, which has affected the elderly rather less than other groups.

Recently, mortgage interest relief has been criticised as a blatant subsidy and there have been calls for either its complete abolition or the restriction of the relief to standard rate tax. General arguments for abolition have also been made on the grounds that the present benefits to any one individual are fairly small, that house prices would not be particularly affected, and that abolition is consistent with the changes that have occurred over the years whereby mortgage interest relief is now virtually the only non-personal relief available to individuals and is, in any case, limited to loans of up to £30,000.

There have been a number of organisations and bodies which have called for the abolition of mortgage interest relief. The Royal Institution of Chartered Surveyors (1986) has argued that tax relief on mortgages should be phased out together with all other help with house costs. The Institution has recommended that the current forms of help with house purchase should be replaced by a housing allowance based on need which would be irrespective of ownership. It would be financed by the £5b that could be 'saved' each year by ending tax relief on mortgages. A report of the National Economic Development Office has recommended the end of higher rate mortgage tax relief to release more resources to pay for repairs to Britain's ageing housing stock. Finally, the *Inquiry into British housing* (1985) chaired by the Duke of Edinburgh has also recommended the abolition of mortgage interest tax relief for owner-occupiers (as well as housing benefit for tenants and the housing ingredient in supplementary benefit) to finance the introduction of a 'needs-related housing allowance' for all those with low or limited incomes to enable them to pay for the housing they require.

It is worth examining the arguments that have been advanced for the abolition of mortgage interest relief or, at least, the end of higher-rate mortgage tax relief.

First, it is argued that exemption of tax on imputed income and capital gains distorts investment decisions. Other forms of investment which attract tax relief are subject to tax on income or capital gains. A landlord of a residential property is liable for tax on rental income and capital gains but is permitted to offset expenses, including interest payments on mortgage loans, against tax liability. By contrast, the argument goes, the owner-occupier receives interest relief but in fact pays no tax on any imputed income and is exempt from taxation on realised gains.

Second, it is argued that the 'cost' of mortgage interest relief has escalated alarmingly. Table 6.1 shows that in 1985–6, the 'cost' to the Exchequer of mortgage interest relief was estimated at £4,750m, equivalent to over 3p off the basic income tax rate. Moreover, it is claimed, the greatest benefit from mortgage interest relief goes to those with the largest mortgages and to those with the highest marginal tax rate. Table 6.2 shows that in 1983–4, 8½ per cent of those receiving mortgage interest relief were on incomes of £20,000 or above, and they received 19 per cent of the total relief. That

Table 6.1: Growth of mortgage interest tax relief, 1979–86

	Total relief	Average relief per mortgage.	Higher rate taxpayers receiving relief
	£ billion	£	'000s
1979–80	1.5	200	80
1980–1	2.0	300	115
1981–2	2.0	450	170
1982–3	2.2	390	150
1983–4	2.8	375	140
1984–5	3.5	410	160
1985–6	4.8	n.a.	n.a.

Source: *Inquiry into British housing* (1985), National Federation of Housing Associations, London

Table 6.2: Mortgage interest tax relief, 1983–4 (by income bands)

	Taxpayers receiving relief		Total relief		Average relief per mortgagor
Income band	'000s	%	£ million	%	£
£20,000 and over	495	8.5	412	19.0	830
£15,000–19,999	760	13.1	353	16.2	465
£10,000–14,999	1,950	33.5	745	34.3	380
£ 5,000– 9,999	2,255	38.7	637	29.3	280
Under £5,000	360	6.2	26	1.2	70
Total	5,820	100	2,173	100	370

Source: *Hansard*, 3 February 1984

assistance is broadly regressive is not surprising, but it is a consequence of the steeply progressive income tax structure in the UK. Moreover, the benefits of relief are spread thinly. Building society figures suggest an average payout from the Inland Revenue of about £660 per borrower per year at current rates of interest. In 1984-5 the average was about £410. When interest rates rise, so does the tax relief.

A further line of argument is that mortgage interest relief has never either helped those who could be described in most need of housing assistance or first-time buyers, many of whom will have relatively low incomes. Moreover, owner-occupiers in receipt of supplementary benefit may actually have their mortgage interest payments met. In 1983-4, about £150m was paid out by the Department of Health and Social Security. The £30,000 ceiling is described simply as a sop to those who oppose mortgage interest relief on principle. Certainly this limit has no logical justification.[1] Since the original £25,000 ceiling was introduced in 1974, prices have multiplied three times. Should the limit be raised to £75,000 or eliminated? A possible worry is that the ceiling has concentrated demand more than otherwise on mortgages of less than or not much over £30,000, and possibly distorted the prices of cheaper houses.

The current nature and extent of housing concessions is alleged to sustain house prices at higher levels than would otherwise be the case. The effect on house prices is believed to be transmitted to other housing sectors, since higher prices on property with vacant possession make it more attractive for private landlords to sell for owner-occupation, rather than reletting under rent control.

Finally, it is argued that mortgage interest relief does little, if anything, either to stimulate new housing or to encourage improvement and modernisation. Indeed, the vast majority of mortgage interest relief goes to purchasers of second-hand houses. Such subsidies as exist to aid modernisation are improvement and repair grants; eligibility for these depends on a property's characteristics, rather than the income of its owners. It is asserted that much of the funds raised through mortgages do not, in fact, finance housing improvements, but appear to leak out on to other consumer durable and consumption goods.

Rather than dissaving or borrowing from other sources, people find it financially attractive to borrow on the security of their homes to obtain tax concessions from the Exchequer. Table 6.3 shows that in 1984, according to Bank of England estimates, some £7b may have been withdrawn from the housing market. It is, therefore, argued that these tax concessions do little to stimulate housing provision, other than raise prices, but appear to provide an important injection into the other private investment markets. Some of the sales of houses are by executors of estates. Since 63 per cent of households are owner-occupied, then it follows that in the course of time a large number of people are going to inherit a house or part share thereof. It may or may not be as good a house as a beneficiary lives in or a useful inheritance, but the proceeds of sale will enable people to become owner-occupiers if they are not so already, or to upgrade their houses if they are.

Finally, detractors of mortgage interest tax relief consider that

Table 6.3: 'Leakage' from the private housing market, 1979–84 (estimates of net cash withdrawals)

£ million

	New new advances	Net private spending on housing	Net cash withdrawal
	A	B	A − B
1979	6,460	4,920	1,540
1980	7,330	6,450	880
1981	9,490	7,130	2,360
1982	14,150	8,430	5,720
1983	14,410	8,680	5,740
1984	16,570	9,360	7,210

Source: *Bank of England Quarterly Bulletin*, March 1985

there is little sound economic reason for limiting personal income tax relief to home loans and denying it to borrowing for other purposes, including financial assets and consumer credit. Moreover, they insist, there is no particular economic merit in owner-occupation, so this form of tenure is not in itself a valid justification for mortgage interest relief. Under present arrangements, therefore, mortgage interest relief is inconsistent, distortionary and has made mortgages a cheap method of borrowing for non-housing purposes and it should, therefore, be abolished.

Some seven million taxpayers have mortgages, the great majority of whom are getting mortgage interest relief. Those who do not are unemployed owner-occupiers whose mortgage interest and other housing costs are met for them. Since autumn 1986, unemployed home-owners under 60 have received benefit to cover half, rather than the whole, of the mortgage interest due on their home loan repayments. After six months, full benefit is restored. Those mortgagors under the Mortgage Interest Relief at Source (MIRAS) scheme are eligible for income tax relief but actually makes interest payments effectively net of 'relief' at the basic rate of tax. Taxpayers on higher rates than 27 per cent receiving mortgage interest relief are estimated at 600,000, or about nine per cent of the total. There are also estimated to be about nine per cent of all mortgages over the £30,000 ceiling, a percentage which increases yearly given a fixed ceiling and rising average mortgages.

As Table 6.4 shows, mortgage interest relief is by no means

91

Table 6.4: Personal income tax deductibility in selected countries

	Owner-occupied housing			Consumer credit
	Imputed income taxable	Tax deductibility of interest	Principal payments also tax deductible	
Australia				
Austria		X	X	
Belgium	X	X	X	
Canada		X	X	
Denmark	X	X		X
France		X		
West Germany	X	X		
Ireland		X		
Italy	X	X		
Japan		X	X	
Netherlands	X	X		X
Norway	X	X		X
Sweden	X	X		X
Switzerland	X	X		X
United States		X		X
United Kingdom		X		

Source: Lloyds Bank and *Taxation, inflation and interest rates* (1984) ed. Vito Tanzi, IMF, Washington, p. 59

restricted to UK mortgages. According to the International Monetary Fund, only Australia among the OECD countries has no mortgage interest relief. Ten countries, including the USA, France, Germany, Italy and the Netherlands have mortgage interest relief, while Japan and Canada, among others actually allow deduction for tax purposes of interest and principal on mortgage payments. Moreover, among those countries that grant mortgage interest relief, six allow deductions for consumer credit as well.

While those arguments discussed earlier are frequently cited as reasons for an early abolition of mortgage interest relief, they fail to recognise both the evolution and status of tax relief within the tax system. Taxation relief on mortgage interest is not a subsidy. The idea that it is arises from its incorporation in the housing policies of successive governments and as a response or defence to the accusations that subsidies to council house building had become excessive. Indeed the various reports cited earlier do not refute the logic of mortgage interest relief but justify its abolition either as a way of 'funding resources' to pay for the 'nation's crumbling housing' or to pay for some new income support scheme.

In fact, there are good reasons to believe that the abolition of mortgage interest tax relief would be inequitable and inadvisable.

The first group of arguments justify relief retention mainly on financial and practical grounds. Mortgage-holders have taken a commitment on a long-term basis and their broad expectations about housing costs should not be sharply disrupted. Withdrawal would probably cause a widespread reduction in consumer spending as mortgage holders quickly adjusted their expenditure patterns to make higher payments. Moreover, if the abolition of relief was restricted to new mortgages there would be charges of unfairness from prospective mortgagees.

Abolition of relief would also lead to objections that loans in the private housing sector are effectively being taxed while housing in the public sector continues to be subsidised. There would probably be a not inconsiderable effect on house prices. The increased effective price of a loan would, to some extent, drive down demand as well as having wider consequences in the money market.

A second group of arguments for retention have a theoretical foundation in that abolition would be a further, and very substantial, move away from fiscal neutrality in personal taxation. Taxes fall on activities rather than people or things. People do not pay tax for who they are but for what they are engaged in. Receipt of income generates a liability to be taxed. There are exceptions, e.g. charities, which do not pay tax on income received. But this is not a subsidy any more than cyclists are subsidised because they do not pay a road fund licence.

Claims that the tax relief constitutes a subsidy arise from government intervention and taxation distortions in 1963, 1967 and 1974. In 1963, Schedule A income taxation on the imputed income derived from owner-occupation was abolished. The system had, in any case, become something of a farce because of a lack of revaluations. In 1967 the option mortgage scheme was introduced, and in the Finance Act of 1974 all tax relief on personal borrowings was removed, except in the case of loans associated with owner-occupied housing and improvements. Currently, owner-occupiers receive tax relief on interest payments on the first £30,000 of mortgage.

TAXATION OF IMPUTED INCOME

Before 1963, individuals who borrowed money for home investment were able to set the interest cost (plus repairs, *etc*) against the liability to Schedule A income tax on the imputed income (housing services) derived from living in their house. In theory, this was a notional market rent the owner received from himself as a tenant and was thus treated for tax purposes as both landlord and tenant. The basis for this taxation is analogous to the business whose income derived from producing a good may be offset against the cost of producing that good (including interest on borrowing) for tax purposes. In other words, payments of interest by businesses are clearly costs of production to be deducted from income before assessment for tax. Under the Income Tax Act of 1952, any person obliged to make annual interest payments and making them wholly out of profits or other taxable gains would be taxed as if he had no such payment, but he would be empowered to deduct tax from the payment and to retain it. Hence, while the recipient is taxed on the interest, the payer is effectively given tax relief. On the other hand, if the interest was not paid wholly out of taxable income, the borrower was obliged to deduct tax before payment and to remit it to the Inland Revenue. It is justifiable that a taxpayer who uses his income to pay interest on a loan should be entitled to reduce his income by a like amount for tax purposes.

It is clear that the existence of Schedule A income tax on imputed income meant that the deduction of mortgage interest for tax purposes did not represent any special subsidy to owner-occupiers at all. The principle was straightforward. Any lending and borrowing was an untaxed activity even though the receipt of income (in kind in the case of owner-occupation) was taxed, provided that the interest was paid out of taxable income rather than out of capital. So long as the imputed income from ownership was taxed, then it was quite reasonable on equity and neutrality grounds to regard interest payments on a mortgage as expenses incurred in the earning of an imputed income. With the abolition of Schedule A in 1963, on the grounds, according to the budget speech of the then Chancellor, of its imperfect base in view of the 1936–7 rating valuations prevailing at that time, this relationship was lost. (In the case of personal borrowings for the purchase of stocks and shares an individual could, at least until 1974, offset the interest and other costs against the income derived in the form of interest or dividends received.)

In the budget speech in 1963 the Chancellor commented, 'It is, however, obvious that we would not charge owner-occupiers of residential property with Schedule A income tax on the new rating valuations. We would then be suddenly trebling or quadrupling the burden of tax on many of those who pay it. This would be intolerable.'

While the political consequences of such revaluation and tax imposition may have been unpleasant, it does not deny the economic logic of the tax or indeed the fact that the abolition has been used as the justification for arguing that owner-occupiers receive subsidies. Moreover, it has also kindled the fires beneath the whole controversy of the relative value of 'subsidies' to owner-occupiers and council house tenants.

There are, therefore, strong grounds on both equity and overseas practice for not only the retention of mortgage interest relief, but the reintroduction of personal tax relief on all loans. Moreover, a tax on imputed income should be reimposed. Although Britain removed it in 1963, other developed countries continue to tax this benefit. Table 6.4 shows that half of the 16 countries tax imputed income from owner-occupation.

The validity of the case for a tax on the imputed income derived from house ownership clearly rests on the acceptance that house investment is a *bona fide* investment and not consumption. In any case, consumer durables are mostly less durable than houses, so a large amount of depreciation would have to be allowed for. For most households, the value of their house is far greater than all consumer durables combined. More practically, it is difficult to calculate the imputed or real income derived from consumption assets, *e.g.* consumer durables. Until the abolition of tax relief in 1974 on all borrowings, personal consumption and investment goods qualified equally for tax relief (in excess of £35 p.a.), but the abolition meant that investments, such as stocks and shares, are now effectively treated as consumption goods.

However, it is not easy to convince an owner-occupier of the validity of a tax on imputed income. After having purchased a house, he finds it difficult to understand why he should be paying a tax on the services it provides. Notwithstanding these aspects, there is an economic rationale for the reintroduction of Schedule A on grounds of tax neutrality and equity principles. Clearly, in view of the upset to people's reasonable anticipations if Schedule A were reintroduced, it would be necessary to consider making the changes in phases, *e.g.* of 20 per cent a year over five years, to allow people time to adapt. On a more mundane level, such a restoration of Schedule A would, of course, require a thorough restructuring of the present rating system which, in itself, is a specific and very significant tax on the consumption of housing.

As a result of the abolition of Schedule A tax it is worth now examining other alleged effects. First, it is frequently argued that tax exemption on imputed income benefits the rich more than other groups because the higher the marginal tax rate, the greater the value of tax relief on mortgage interest. This is a *non sequitur*. Whether or not mortgage interest relief exists, the rich will be tempted to house themselves more lavishly than they would if the imputed income was taxed. While it is true that the higher the marginal rate of tax of the borrower, the lower the net interest rate on the mortgage, this is inevitable given the progressive tax rates in Britain. Moreover, the £30,000 limit means that very often at the margin higher rate taxpayers don't get any additional tax relief if they borrow to improve their housing.

Second, it is argued that tax exemption discourages building for rent. Clearly there have been many adverse factors to discourage house-building for letting in the post-war period but, in addition, taxation in respect of private letting has disadvantages. First, for tax purposes, houses are assumed to last for ever and landlords may not deduct any depreciation allowance from rents received before tax assessment. Second, since 1974, mortgage interest on loans for the acquisition of or improvement of houses to let may only be set against rents received from the property and not against the overall income of the landlord. However the potentially distortionary effects, as a result of the Finance Act 1974, are far more pronounced in the case of 'non qualifying' personal loans, *i.e.* most loans not for the purpose of house purchase or improvement. As explained earlier, the person who now borrows money to invest in company securities receives no tax relief against interest or dividends received.

Third, it is claimed that tax exemption means that tax relief is an open-ended subsidy. Although the total relief on mortgage interest has increased considerably in recent years, £865m in 1975-6 and estimated at £4,750m for 1985-6, this increase is largely attributable to higher tax rates and higher mortgage rates. Tax relief cannot be regarded as open-ended so far as the Exchequer is concerned, since interest paid to depositors is taxed under the composite tax arrangements; there are therefore strong theoretical grounds for saying that tax relief is not a subsidy. This is because as the tax rate rises the tax relief will also increase, but the shortfall is largely made up automatically through the increased composite tax levied on shares and deposits of building societies. Similarly, if the share rate rises with a corresponding increase in mortgage rate to maintain the margin, the increase in tax paid on savings balances will be similar to the increase in the value of the tax relief on mortgage interest. Thus the factors which cause tax relief to increase must inevitably cause income tax collected to be increased by a similar amount. The relationship between tax relief on mortgage interest and tax paid on building society shares and deposits is important. While the increase in tax relief and composite tax rate go hand in hand, it could be argued that the determination of the composite tax rate, defined as the average marginal liability to basic rate tax, implied a special type of subsidy from one group of savers to another. It is a taxation subsidy from those savers whose incomes are too low to bring them within the taxation net to those whose marginal rate is the basic rate. However, while it is true that people not liable to income tax are unable to claim the income tax paid on their behalf on building society interest, the obvious course of action if they object is for them to withdraw their deposit and invest it with an institution which pays interest gross. For those investors liable to tax at higher rates, this has to be paid separately.

Finally, there is the popular notion, discussed earlier in the chapter that the 'net cash withdrawal' from the private housing market has fuelled a massive consumer boom. The 'net cash withdrawal' is the difference between net new loans for house purchase and net private expenditure on housing. While the Bank of England has estimated the total at £7.2b for 1984, equivalent to three per cent of total consumer spending, it is wrong to jump to the conclusion that 50 per cent or all is actually used for domestic consumption thus fuelling a spending spree. It is likely that only a very small proportion of 'net cash withdrawal' is diverted into consumption. More is likely to go directly back into the financial system itself to sustain the volume of new mortgage lending. There are a number of sources of 'net cash withdrawal'. However, these 'sources' are frequently negated by other economic behaviour. For example, sales of rented property may initially create a 'cash withdrawal' but this will be offset by the mortgages of those who

97

buy the property. An owner who re-mortgages may provide for himself a source of 'cash withdrawal' but mortgage lenders generally only authorise such increases in mortgage on evidence of housing work done. Similarly, proceeds of house sales of those who die provide a source of cash withdrawal, but there is no real evidence that the beneficiaries of proceeds go on a spending spree. Much of the money, as the Building Societies Association points out, goes back or to them as deposits. Alternatively a considerable proportion is likely to be invested in securities or property.

CAPITAL GAINS TAX EXEMPTION

In contrast to other investment assets, capital gains arising from the sale of a principal place of residence are exempt from capital gains tax. This tax is by no means applied consistently and depends on the nature of the asset. In theory, this exemption may be described as a subsidy but the issue is more complex than at first sight because of the nature of the gain. The majority of capital gains to owner-occupiers are holding gains as opposed to operating gains. In other words, the gain during periods of inflation is a paper gain rather than a 'real' gain. Since the Finance Act 1982, the purely inflationary part of capital gains is no longer taxed. If a number of sellers are using the proceeds to buy other houses whose prices have also risen, taxing capital gains on housing would have adverse effects on mobility. House prices have generally increased in money terms by the same amount and the 'real' capital gain on individual houses has been only a small proportion of the 'money' capital gain. This has had a beneficial effect on landlords in respect of the sale of houses which are not the principal residence of the owner. Capital gains tax is now levied on the real increase in the value of the property and not the nominal gain.

While it is true that the capital gains tax exemption is a relative advantage to investment in owner-occupied housing, broader reforms are implied through the application of either consistent treatment among all investment assets for capital gains tax purposes or its abolition.

Income distribution effects of taxing imputed income from owner-occupied housing

If owner-occupied housing was added to the tax base but tax rates were held constant, there would be an enormous increase in government revenue. This however would allow a rise in tax thresholds and/or cut in tax rates. Most taxpayers would thus find their net tax position changed very little. People would gain if they were not owner-occupiers but paid tax; however a large proportion of tenants are subsidised in various ways, and the subsidy schemes could if desired be adjusted to claw back their tax savings. People would lose if their housing stock was high relative to their income; this might particularly affect the retired, and again allowances could be changed to compensate. If everybody was an owner-occupier and house value was linearly related to income, then if the old tax system was linear, a new linear system could be devised which left everybody's tax unchanged if they kept the same house. Clearly the actual position is more complicated than this and there would be some changes in net tax position.

If the result of taxing owner-occupied housing was a more efficient use of the country's housing stock, and in the longer run of the country's labour force, this would give overall gains, some of which could be used to compensate tax losers.

Suppose that with no taxation of owner-occupied housing the tax system is described by:

$$T_0 = A + bY.$$

Use of owner-occupied housing services is given by:

$$H = C + dY.$$

Suppose the tax base is changed to tax-imputed rentals; thus

$$Y^* = Y + H.$$

Then the new tax system is:

$$T_1 = E + fY^*.$$

It is possible to set E and f so as to make their tax position the same for every individual:

$$T_1 = E + f(Y + C + dY) = (E + fC) + (f + fd)Y.$$

For tax paid to be unchanged for everybody we need $T_1 = T_0$ for every Y, so

$$E + fC = A, \quad i.e. \quad E = A - fC$$

and

$$f(1 + d) = b \quad i.e. \quad f = b/(1 + d).$$

The real world position will be more complicated since T_0 and H are not in fact as described.

NOTES

1. There is also the anomaly whereby a married couple living together get only £30,000 worth of mortgage relief, while an unmarried couple can get £30,000 each. There is a quite hefty tax on being married if the mortgage is over £30,000.

7

Reform of Council Housing

THE ISSUES

During most of the post-war period, the building, financing and allocation of council housing has manifested an unacceptable degree of monopoly renting power and bureaucratic inefficiency and incompetence. With the great majority of decisions taken by suppliers of housing services comprising central and local government politicians and officers, with little regard to tenant interests, the results have been appalling. Houses and flats have been built to poor designs, vast estates have emerged, rents have been kept artificially low to create and stimulate shortage and houses have been allocated by political patronage. It is amazing that houses and flats built only 20 years ago at high cost and poor quality are having to be demolished not only because of their appalling condition but because tenants refuse to live in them. Moreover, the momentum is gathering pace as the legacy of 60 years of council housing is one of a vast increase in public renting displacing private renting and, finally, of council housing itself being gradually displaced as sales and transfers of council houses gather pace. Housing segregation, immobility of tenants and rising rent arrears have become the products of a succession of ill-founded past decisions in the name of social engineering, planning, development and civic blunders.

Perhaps the greatest issue in council housing policy and finance has been the persistent desire of all councils to keep rents excessively low, often subsidising rents from rate funds let alone from Exchequer subsidies. Keeping rents below replacement or economic values has resulted in excess demand for council housing with organisations such as Shelter frequently claiming that a million applicants are on council housing waiting lists. Rents bear no relation to market conditions or even what tenants are willing to pay for a dwelling. Indeed, the absence of any signals from the market place on price, quantity and quality of housing inevitably leads to the problems of misallocation and disaffection between suppliers and consumers. Moreover, the waiting list becomes the single, but wholly imperfect, indicator of apparent housing need. The lower the rent levels the higher the waiting list as people register from other areas or from higher-cost private letting, and the greater the apparent housing need.

Until fairly recently, rental agreements between local authorities and their tenants were notoriously unfair to tenants. Agreements were either vague or almost absent at best or were downright restrictive at worst. Not for nothing were local authorities accused of bureaucratic hegemony in the restrictive clauses of agreements. Perhaps worst, these agreements were honoured by most citizens at considerable cost to themselves in the absence of choice and freedom of their own home; but in the case of some tenants who abused their home and surroundings, local councils failed to take effective action to uphold rental agreements.

In the face of present and cumulative expenditure on council houses some attempt has been made to sell council houses to sitting tenants. It was not until the Conservative government was elected in 1979 that sales started to become significant. In 1980, fewer than 100,000 houses were sold to sitting tenants but in both 1982 and 1983 this increased to 200,000. By 1983, publicly owned land and housing worth £1.5b annually was being sold, of which £1b was council housing bought by tenants under the Housing Act 1980.

In the early 1980s, this policy had some success, but in the current climate of low council house rents, and high real mortgage interest rates, the incentive to buy is less evident in spite of the considerable discounts available. At one stage, such sales were highly controversial; they were opposed by most Labour-controlled authorities in England. Today, the popularity of such sales has led to some rather muted opposition to council house sales from the Labour party.

METHODS OF PRIVATISATION

Council sales

The Housing Act 1980, introduced right-to-buy terms whereby tenants were to have the right to not only buy their homes but also, if they wished, to do so through a council mortgage. The purchase price payable by a tenant was to be equal to the market value of the property, less a discount.

Under the 1980 Housing Act, tenants of between three and four years' tenancy were to receive a discount of 33 per cent. If the period of the tenancy exceeded four years, then the discount rose by one per cent for each additional year, to a maximum of 50 per cent for tenancies of 20 years or more. The act included other 'rights', such as the depositing of £100 with the local council and thereby freezing the valuation placed on the house for up to two years if a tenant did not wish to purchase the house immediately. The actual discounted price for a council house could not be less than the costs incurred by the council in the building of the dwelling.

Following the fillip given by the improved terms under the Housing Act 1984, the government expected sales of 200,000 in that year. In fact, only 150,000 sales were effected at an average discount on sales by authorities of 45 per cent. By the end of 1984, Table 7.1 shows that some 530,000 homes had been sold by local authorities and new towns in Britain under right-to-buy legislation. In 1985, another 100,000 sales were effected. Overall, about 95 per cent of the sales have been to sitting tenants and about 94 per cent of sales of local authority and new town dwellings in Great Britain in 1984 were of houses rather than flats or maisonettes.

Equity sharing schemes

Joint tenure schemes, similar to those pioneered by the Birmingham City Council, are now promoted by many councils and have attracted far greater political support than council house sales. Under these schemes an alternative form of tenure is provided to those who may wish to purchase a house but whose income is inadequate for an average mortgage. Under a half-and-half scheme, the purchaser may borrow half the value of the house, through a local authority mortgage, but continue to pay rent on the remaining half.

Table 7.1: Sales of dwellings owned by local authorities and new towns, 1971–84, England and Wales; and 1980–4, United Kingdom[a]

'000s

	Local authorities	New towns	Total sales
England and Wales			
1971	17.2	3.1	20.3
1976	5.8	0.1	5.9
1977	13.0	0.4	13.4
1978	30.0	0.6	30.6
1979	41.7	0.8	42.5
1980	81.5	4.2	85.7
1981	102.3	3.7	106.0
1982	201.9	5.2	207.1
1983	141.6	4.9	146.5
1984	102.6	4.3	106.9
United Kingdom			
1980	86.1	6.0	92.1
1981	118.8	5.5	124.3
1982	221.3	6.4	227.7
1983	163.0	7.2	170.2
1984	122.8	6.3	129.1

Note: a. Including leases, sales to housing associations, and sales of dwellings previously municipalised.
Source: Department of the Environment

At any future date, the purchaser may acquire the remaining half share, subject to valuation at that time of the increased value of the council's half share, less the cost of improvements. In the interim period, the rent of the council's half share is charged as half a normal 'reasonable' rent, normally subject to the rent being at least two-thirds of the rateable value in order to prevent difficulties under the Leasehold Reform Act 1967. These schemes provide some flexibility for both the tenant and the authority. Since preference is normally given to existing council tenants, it does provide a house for immediate letting. The major practical problem for the attraction of owner-occupation either by local authority house sales or shared equity schemes lies in the relative cost of a rent and a mortgage.

Equity sharing and right-to-buy legislation have had a profound effect on the choices available to council tenants. Notwithstanding the actual sales to date, over 25 per cent of UK housing still remains within the administrative and political control of local authorities. More opportunities need to be offered to enable a greater transfer .of council housing to independent local bodies offering greater flexibility, mobility and tenant participation. Since central government financing of local authority housing is likely to fall further in future years, government should enhance its present right-to-buy policies with new initiatives requiring local authorities to relinquish their funding and management of local housing.

Devolution of council house management to tenants' cooperatives and trusts

Management devolution of council estates to cooperatives and trusts is being effected through grouping the housing stock into small natural communities by the establishment of co-ownership cooperatives to take them over. Tenants are able to have a direct stake in the house or building they occupy and so ensure responsiveness by the management to tenants' needs and desires. Under a cooperative management arrangement, tenants have a direct stake in improving the value of their house which can be realised upon sale or used as collateral for loans. The form and financial structure of cooperatives vary considerably but key principles are enshrined in legislation for the devolution of council estates and parts thereof to cooperatives, *e.g.* one share, one vote in respect of each housing unit, the right to sell the share, *etc*. Moreover, comprehensive arrangements can be incorporated in cooperative articles of association.

Sales and transfers of empty and undesirable houses and estates

In council housing estates which are hard to let, sales of empty council houses to private developers to refurbish for sale or rent offer another initiative and would provide a new refurbishment opportunity. The Department of the Environment has already launched an Urban Housing Renewal Unit to tackle the problem of the 1.3 million sub-standard homes on run-down and badly managed council estates. The unit is intended to work with local authorities to provide assistance and draw in new private-sector funds and urban development grants to supplement existing housing investment

105

programme (HIP) funds from the department. This initiative emerges because of the scandal of so many recently built and structurally sound dwellings falling into decay and unpopularity when houses are standing empty, even though they may be situated in areas where large numbers of families are living in expensive bed and breakfast accommodation or in overcrowded or sub-standard conditions.

The government has good grounds to push ahead with privatisation by selling estates to developers and building societies. After refurbishment and rehabilitation the houses can be sold as low-cost housing. Building societies are able to make such schemes self-financing by selling some of the homes and raising rents of the remainder to economic levels.

A number of housing schemes have been initiated. For example the scheme at Thamesmead offers, to date, the most ambitious opportunity. Upon the abolition of the Greater London Council, residents voted to manage their own estate. The Building and Planning Act 1986, was intended to permit more councils to sell entire estates to tenants' trusts or cooperatives. As discussed earlier, new legislation offers an element of compulsion on local authorities at least to seek out the wishes of tenants through some formal machinery. Thamesmead chose to manage their own estate through a number of districts, each electing a representative. Given its enormous overall size, Thamesmead is an extreme demonstration of the possibilities of such cooperation.

LIMITATIONS TO COUNCIL HOUSE SALES

There have been a number of arguments put forward by antagonists of council house sales, and their validity should be examined:

'The sale of council houses reduces the pool of houses and jeopardises re-lets'

The effect of sales on the actual stock of local authority housing is frequently subject to some confusion. While it is true that there is an immediate transfer of tenure upon sale, it is not true that there is a corresponding loss to the available letting pool and hence that needy applicants are denied accommodation. This is because the average vacancy rate of local authority housing is only about four per cent per annum, reflecting the low mobility of council tenants. Thus for every 100 houses sold by a local authority, there is a presumption that the actual loss of re-lets is only about four houses since the other 96 new owners would have remained as council tenants had they not purchased their houses. This argument clearly presumes that both the tenants who buy and the type, style and location of houses sold are typical of the council house market.

That re-lets are jeopardised by sales is difficult to justify for it means arguing from very insignificant and unrepresentative cases and is highly speculative. Recent surveys of sales of council houses make clear that the most typical tenants likely to purchase their council homes are well-established people in middle age with a considerable number of years ahead of them before their households would be dissolved either by death or by going to live as part of someone else's household or in an old people's home. There is little reason to suggest that these tenants would have moved away to buy if they had not had the opportunity to buy as sitting tenants at a concessionary price. Until vacancies due to death become numerous, there will be very little effect on the number of re-lets.

Of course, some of those who have purchased their homes might have been sufficiently keen on becoming owner-occupiers to have bought something else if their houses had not been available for purchase. If there was a rise in rent levels towards economic rents there would be more such cases. So sales to sitting tenants probably do reduce the vacancy rate.

'Only the better houses will be sold'

There is little evidence to support this contention. The attraction of a house is reflected in its price and council houses of all types and quality have been sold in recent years. That relatively few flats have been sold (30 per cent of all tenants live in flats) owes more to the legal difficulties and delays in negotiating sales than to a lack of demand at a market price. Moreover, it has been evident throughout England and Wales that where poor quality housing has been offered for sale it has commanded widespread demand. Homesteading schemes and other initiatives have demonstrated the interest of people in acquiring blighted and poor quality housing for improvement. By contrast, people will also be keen to buy the best council houses, if only because it is likely that those who want and can afford their houses will already have done most to improve them as tenants.

Moreover, the overall quality of the estate will determine sales and price levels. Where sales have been successful, the remaining council tenants are surely beneficiaries of any improvement by owners who were previously tenants.

107

'Substantial financial losses are incurred by the public sector from the sale of council houses'

It is frequently alleged that the legal and high levels of discount on the valuation of the house (on the condition that the net selling price is equal to or greater than the original cost of the house) is too favourable to prospective purchasers and means that public assets are being given away. This argument ignores the fact that such discounts on sales include a pre-emption clause limiting resale before a specific number of years. Moreover, the sales are to sitting tenants who are, in any case, protected tenants. For these reasons, some discount is clearly appropriate, but, of course, it limits capital receipts to local authorities.

In respect of the financial effects, the tenant and local authority are affected according to the source of finance for the purchase, and it by no means implies a reduction in public expenditure through capital debt redemption by authorities, or enables local authorities to build additional houses with the proceeds.[1] If sales are financed directly by the tenant for cash or by building society mortgages the local authority capital debt can be reduced. Few tenants are in a position to pay cash for their houses, or even put down substantial deposits, and the building societies may be reluctant to finance such purchases for such reasons as a prospective mortgagor's non-membership of a society, the valuation basis and possibly the pre-emption rule. At the beginning of the 1980s the great majority of sales were financed through local-authority mortgages. In recent years, however, a reducing number of sales are being financed from council mortgages. Under a local authority mortgage the impact on the housing revenue account varies according to the debt outstanding.

The Department of the Environment (1980) has shown that for the first 20 years, under all assumptions, the sale of council houses should be financially beneficial both to local authorities and to the Exchequer. Beyond that time, a 20-year projection of the effects is more speculative but, according to the government, sales should bring benefits not only to council tenants but to the community as a whole. The Select Committee on the Environment (Session 1980–1, HC366) on Council House Sales was unable to refute this proposition, notwithstanding the evidence it heard to the contrary. Even then, the Committee was able to say only that long-term financial effects depended upon assumption about future costs, benefits, payments and receipts and that no clear prediction could be made. In short, even on financial grounds, where long-term losses may occur, this is no reason to deplore sales now where the short-term benefits are not only favourable but may actually exceed any longer-term losses.

'Council house sales stigmatise remaining tenants, forcing them into second rate or even slum housing'

Again it is difficult to follow this line of reasoning, let alone find evidence to support this contention. Perhaps the deductions that prompt this assertion say more for the attitudes of those interventionists who subscribe to this notion. Are its proponents saying that they wish poor people to remain not only indefinitely in rented homes on poor estates but that they should remain wholly dependent on public welfare? If council house tenants actually do have security of tenure for life, then a change in tenure in no way affects the overall availability of housing. Moreover, it does not confine the benefits to those who have become owner-occupiers. It raises the prospect of greater benefits for tenants whether they remain tenants for life or subsequently decide to purchase their home: it is also financially beneficial to local housing authorities.

The arguments against council house sales are manifestly weaker than their antagonists will ever admit. Indeed, it is difficult to find any reasoned justification for councils to hold on to their monopoly over public rented housing and this is evident in recent and new government legislation intended to promote new initiatives to sell off or transfer the ownership and management of council estates.

Successive surveys by the Building Societies Association provide evidence of a massive desire for owner-occupation among all adults. An all-time high of over 60 per cent of all houses in the UK are now owner-occupied. Although the likelihood for sustained and very large sales of council houses to existing tenants appears now less encouraging than the overwhelming desire for owner-occupation *per se*, there are good reasons for expecting this. First, nearly two-thirds of council tenants are on income and rent support of one sort or another and 33 per cent of all tenants are pensioners. Secondly, the responses to questions on the likelihood of council tenants purchasing their own home will have been answered in relation to existing terms. There is still plenty of scope for government to improve terms further and to set prices which will stimulate interest in the case of less desirable properties which because of either their quality or their environment appear less attractive. A number of strategies are available to encourage those who, if the terms are right, would be more amenable to purchase or a change of tenancy arrangements.

The government has made a number of modifications to the terms and conditions for the sale of council houses, including the Housing Act 1984 which improved eligibility and increased discounts. In 1986, the average discount was 42 per cent. Under the 1986 Housing and Planning Acts improved discounts and terms of resale for tenants were permitted. Discounts are offered on a graduated basis according to number of years of occupation. For example, after two years occupation of a house the discount on valuation is 32 per cent, rising to 60 per cent maximum after 30 years. After two years' occupation of a flat the discount on valuation is 44 per cent, rising to 70 per cent maximum after 15 years. These new discounts could have a significant effect: flat sales still account for only four per cent of sales, even though flats make up 33 per cent of total council house stock.

However, there still exists scope for improvement in discounts and these should be prosecuted further, for both short- and long-term tenancies on houses and flats. The right-to-buy could also be extended to the tenants of charitable housing associations.

A further improvement in the Housing and Planning Act concerns conditions of resale. Until now, five years had to elapse before a house could be resold without restriction or financial penalty. A new owner wishing to sell a former council house before five years had elapsed had to pay back part of the discount to the council. For example, a tenant who had purchased for £8,700 a house valued at £15,000 would have had to pay back nearly £4,000 if he moved within three years. There seems little justification for such a rule. Under the new legislation, the government reduced this limit from five to three years. Tenants receive discounts in their capacity as sitting tenants. There is no reason why they should not capitalise on their purchase by an early sale if they so wish. Restrictions slow down activity in the housing market and hamper people's geographical and occupational mobility. This new permission to sell after three years without loss of discount will provide an important relaxation, especially in the case of new home-owners who are not prepared to accept a job in a different area because they would owe a local council too much money.

Ultimately, most council housing could perhaps be taken out of local authority control and transferred to tenants' cooperatives, coupled with the compulsory requirement of sale of difficult-to-let, vacant housing and unused local authority land to individuals and developers. There are currently estimated to be 115,000 empty council houses and flats, with some 30 per cent unoccupied for more than a year, at a time when rent arrears are over £200m. Such a wholesale transfer implies the breaking up of council estates by encouraging tenants, building societies and pension funds to manage them. Under the government's new proposals, tenants will have the right to petition the government for a building society to take over the management of their streets or tower blocks if they think the building society will manage them better.

FUTURE OF COUNCIL HOUSING

Rents in the mid-1980s still only take up ten per cent of the average manual worker's salary, compared with seven per cent in 1979. The Audit Commission, an independent body set up by the government to scrutinise councils' accounts, reported in 1985 that many councils still charged only half of what they needed to cover the cost of properties. The Audit Commission has argued that rents are too low to keep buildings in good repair and that an increase of £2 on the average rent would generate more than £450m a year on the 4.4 million council dwellings in England. In any case, there are no direct penalties for councils that ignore government guidelines that rents should rise in line with inflation. Councils have had considerable freedom to fix their own rents ever since 1975, when the then Labour government repealed the Housing Finance Act of 1972.

In the meantime, council housing costs continue to rise. In the last ten years, management costs have risen two and a half times faster in real terms than maintenance costs. With rent rises so repressed that in some cases they cover only some 30 per cent of costs before debt service, there is no prospect of any real contribution to debt servicing or capital expenditure programmes.

There are an enormous number of legislative barriers to private housing development within inner cities. Large urban sites owned by the private or public sector have remained undeveloped for years. Problems range from local authorities refusing to release suitable building land, to landlocked sites owned by both the private and public sectors. The Urban Development Grant and its successor, the Urban Regeneration Grant, have gone some way with important financial incentives but what is needed is simplified planning zones designated in the inner cities and a reduction of the present planning bureaucracy. Local authorities are frequently able to exceed themselves in justifying the need to maintain enormous land banks, thus blocking new initiatives and commercial development by the private sector.

111

HOUSING ASSOCIATIONS

The status of housing associations was formalised under two acts. Under the Housing Act 1972, the Housing Corporation was empowered to advance loans to housing associations and so augment the level and diversity of traditional sources of loans. The Housing Act 1974 specifically recognised housing associations as quasi-statutory bodies within the public sector, and successive governments have included the expenditure of the associations as part of the total public housing expenditure. At the same time, however, the 1974 Act laid down a formal financing framework for housing associations which considerably reduced their previous autonomy.

Some housing associations have their roots in the last century and they have always cherished their independent and voluntary approach to housing. There are currently several thousand registered housing associations and their objectives remain the housing of those in need, as locally perceived, through the rehabilitation of existing houses and housing developments. Resources of individual housing associations have never been large and while they lack the ability to compete for land and finance *vis-à-vis* the local authorities, the associations have initiated development and growth using wide-ranging public and private sources of finance.

Under the Housing Act 1974, housing associations receive once-and-for-all capital grants to cover, and effectively write off, all loan debt which cannot be funded from expected net rental. The objective behind this capital grant system is to prevent housing associations receiving annual revenue subsidies like local authorities, thus avoiding any open-ended and sustained subsidy commitment.

In contrast to their importance to our European neighbours, housing associations account for a mere two and a half per cent of Britain's homes. Nevertheless, more than one million people in England and Wales live in about 550,000 homes, all of which have been reclaimed from dereliction or built from scratch by more than 2,000 housing associations.

During the 1970s there was a considerable increase in housing association building. In 1968, only 6,000 homes were built. By 1983–4, building and rehabilitation was nearer 20,000 houses, down from over 30,000 in the late 1970s.

Housing associations which are registered with the Housing Corporation receive financial assistance through a combination of loans and capital grants, which enable the associations to charge 'fair rents' set by a rent officer. The actual 'fair rent' set and hence the income received by the housing association must be sufficient to finance the housing project, *i.e.* losses should not be incurred. The initial grant from the Housing Corporation is expected to be repaid in due course and, in theory, is pooled and redistributed to fund new projects in areas of higher priority of need.

The low rents charged by housing associations ensure that not only are they unable to satisfy the growing demand for their services from the young, unemployed, single, disabled and elderly, but they also face the prospect of sustaining huge reversals in their housing programmes because of successive annual declines in funding.

In 1984–5, the Housing Corporation provided approximately £830m of funding and, in addition, local authorities (at 'no cost' to ratepayers or tenants) also provided local support to housing associations. By 1986, that figure had fallen to £660m. In view of cost controls, support from local authorities has also fallen from £456m in 1977–8 to £111m in 1986–7, more than 75 per cent in ten years.

Consequently, the number of new homes that can be provided by housing associations has shrunk drastically. Since 1982–3 they have had to reduce their programmes by 30 per cent in real terms, which has meant a decline from 40,000 to 20,000 new housing projects each year.

In the early 1980s, the Department of the Environment and the Housing Corporation concentrated funding for housing associations in 80 key regions such as Liverpool, Birmingham, Leeds and London, designated as 'stress areas', where the combined effect of urban deprivation, unemployment and racial tension is greatest. As a result of that decision, 83 per cent of government funding for housing associations now goes to stress areas, while the rest of the country is left with 17 per cent of the total budget, a decline of ten per cent compared with earlier years.

Housing associations' tenants, unlike council tenants, do not have the right to buy their own homes. Moves to extend that right to all tenants were included in the 1985 Housing and Building Control Bill, but were abandoned by the government following the defeat of this clause in the House of Lords. However, it is difficult to understand why tenants of accommodation provided by housing associations through the support of public funds are treated differently from other tenants in the public sector.

FUTURE POLICY

The government needs to continue to improve the selling terms of council houses and remove further petty restrictions on resales. Sales promotion needs to be targeted on specific groups with the better-off, young and middle-aged more strongly induced to buy through a combination of improved buying terms on the one hand and a move to (higher) economic rents on the other. Also, people occupying less desirable houses should be offered new and special terms through benefits in kind, cut-price installations and improvements, or cash discounts. Tenants of very undesirable tower block estates should be relocated and these blocks should be sold to private associations for rejuvenation and re-letting at economic rents. Other forms or groupings of property should be transferred to housing associations or cooperatives. Those tenants who chose not to purchase their homes ought to be protected financially through state income, as when they were council tenants. The rents paid, however, would be at economic levels to induce full rejuvenation.

Such policies orientated towards target groups through a continuation of sale and transfer would bring about important social and economic benefits: reduction of social tensions, elimination of petty restrictions, greater geographic and occupational mobility and a demise of the great council bureaucracies.

Unlike any other initiative in housing policy, acceleration of council house sales together with massive transfers to housing associations, building societies, or cooperatives would provide the necessary spearhead towards the notion of a property-owning democracy.

Encouragement of large scale privatisation would improve the quality of housing. Sales offer a sustained and irreversible redistribution of wealth to those who have for so long never been able to contemplate any real choice. Those preferring to rent from either a council or other body should be able to do so, but at economic rent levels.

A commitment to council house sales has not only been one of the few housing policy successes by a government in mastering tax expenditures on welfare but has given an important fillip to home ownership and choice. Annual sales of between one and two hundred thousand per annum may sound impressive, but at present they make hardly a dent in the 5.7 million houses still owned by local authorities.

NOTES

1. Councils are more inclined to retain capital receipts to finance new house building and renovation rather than to redeem debt.

8

Policy Options for Private Rented Housing

In the face of such adverse effects of rent control, what hope exists for a reversal of the worsening trends in private rented housing? Policy proposals to at least protect the residual of rented property have centred on either the complete social ownership of housing through government takeover of responsibility for this housing stock and tenants, or proposals for a revitalisation of the private rented sector ranging from the mild to the radical. Undoubtedly, either total or partial municipalisation or the continuation of rent control requires considerable supporting measures. The real issue, however, is whether the government is to assume direct responsibility for this housing group or whether the private sector is to be provided with a climate conducive to a healthy rented sector unfettered by the current severe restraints of government regulations.

The scope for and implications of the main policy options of municipalisation, expansion of the activities of housing associations and the voluntary sector, reforms under the present general framework of control and, finally, abolition of rent and eviction controls are now examined separately.

MUNICIPALISATION

Such a policy option would necessitate gradual and complete municipalisation of all privately rented property. It would have to be financed by long-term loan issues together with fairly considerable Exchequer subsidies to bring the acquired properties up to local authority housing standards. While the last Labour Government ruled out municipalisation, proponents (Wicks, 1973) continue to argue that municipalisation is the only real solution for the privately

116

rented sector. They argue that the 'housing problem' will only be solved when all rented property is under government or 'community' ownership and entirely outside the ownership and control of private landlords. If the local authorities were to take over all private lettings the costs would be considerable and few local authorities have ever been very enthusiastic. They would assume immediate and total responsibility for dwellings which, under government minimum standards, are inadequate, require massive renovation and possibly demolition, and provide a new home for tenants. Even if central government finance was forthcoming, local authorities would have to take on responsibility for implementation of policy through acquisition and landlord 'compensation', rehousing tenants during renovation and repairs, clearing condemned houses and rehousing families. All this would be in addition to their current pressing obligations. Arguments for local authority takeover of the existing stock have been summarised by Murie *et al* (1976) who have contended that such a policy could provide:

(a) increased control over the sector and occupancy standards, since municipalisation would automatically ensure that the fair rent system and tenancy rights were extended to all existing tenants; and

(b) improved allocation systems through allocation according to 'need' as perceived by local authorities rather than by a subtle combination, according to Murie, or 'rents, search and queueing costs which tended to favour young persons with higher levels of incomes and experience in particular local housing markets'.

Proponents of municipalisation have neither costed their proposals nor critically questioned the reasons and implications of such a bizarre allocation system under a controlled rental system. With so many local authority houses in poor condition it is doubtful whether the supply of improved properties would increase in inner city areas if left to local authorities.

POLICY OPTIONS UNDER CONDITIONS OF RENT CONTROL

In the general framework of rent control are there any policy options that exist which could revive the rented sector? Before any policy options are explored some attempt should be made to assess the theoretical derivation of the notion of a 'fair rent' and its relationship to economic forces. First, it should be recalled that a fair rent is determined according to the attributes of the property and excludes the financial and personal circumstances of both landlord and tenant.

Because the number of persons seeking to become tenants is assumed to be equal to the number of houses for rent, an extreme and obviously unintended economic interpretation would suggest that scarcity is to be ignored and the fair rent set at zero price. However, government intention under this clause was clearly that only 'abnormal' scarcity was to be discounted. Hence fair rent is defined at the level where, in the absence of abnormal circumstances, need would be equated to supply with neither landlord nor tenant penalised.

The model represented in Figure 8.1, developed by Cooper and Stafford (1980) attempts to interpret a 'fair' rent and embraces the following assumptions:

a) Housing units are homogeneous with the same site value and each family can only possess one dwelling unit; b) OS_1 is the actual stock of housing to rent and S_1S_1 is a perfectly inelastic short-run supply curve; c) S_2S_2 is the long-run supply curve assuming a perfectly competitive market, with all firms having the same long-run average cost curves and no technological or pecuniary externalities.

In a perfectly competitive market, P_1 is the 'fair rent' because it would be the rent in the absence of scarcity under constant cost conditions, i.e. the long-run perfectly competitive equilibrium rent level. Thus P_1, the 'fair rent', is defined as the level where, in the absence of abnormal circumstances, need would be equated to supply with neither the landlord nor tenant penalised. Under this determination of fair rent, the government assumes that OT families are both ready and able (at least with assistance through rent allowances) to pay the rent of P_1 for some standard of accommodation of given quality and location. In this context, the rent of P_1 is fair to both landlords and tenants and the demand curve DD will, under these assumptions, have a kink which will coincide with the fair rent P_1. Demand is therefore perfectly inelastic up to and including the fair rent level on the earlier assumption that people do not demand more than one unit.

An initial situation of disequilibrium is assumed in the model with P_1 rent prevailing as the fair rent. Wnile OT families are ready and willing to pay P_1, supply in the short run is only OS_1. Therefore, in the short run at least, equilibrium can only be achieved at a rent of P_2. In the long run, assuming a perfectly competitive market, new entrants would ensure that a rent of P_1 would ultimately be attained and OT families housed.

In Figure 8.2, the more realistic assumption that long-run supply

Figure 8.1: What is a fair rent?

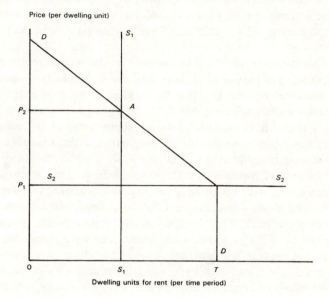

Price (per dwelling unit)

Dwelling units for rent (per time period)

Figure 8.2: Fair rents with increasing costs

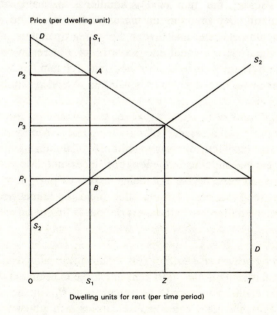

Price (per dwelling unit)

Dwelling units for rent (per time period)

is subject to increasing costs is made and the supply curve S_2S_2 slopes upwards from left to right. Since no obvious long-run solution presents itself, given the control of rent of P_1, the state is faced with a number of possibilities — but they all pose difficulties:

A Control rents at P_1. Rent control at P_1 would represent a perpetuation of the policy of 'fair rents' and theoretically implies the transference of P_1BAP_2 suppliers' surplus to consumers and prevents them 'exploiting' the short-run shortage. While the rent is fair in that it fully meets the long-run supply price of OS_1 units, it in no way provides an incentive to suppliers to increase supply, even in the long run. Moreover, it is potentially unstable, for landlords are well aware that there are OS_1 people willing to pay a rent of P_2 or more, and will, accordingly, attempt to circumvent rent legislation. Moreover, once the rent is controlled at P_1, it will be politically difficult to remove the control, for there will be an excess demand of S_1T (OT families are willing and able to pay the fair rent) and a queue will emerge.

In the case of those landlords who attempt to evade the law, some of the enhanced consumer surplus will find its way back to them. A black market will emerge with such practices as key money and payment for fixtures and fittings. Some families are even willing to pay in excess of a rent of P_2. Landlords will also attempt to increase utility by rationing on prejudice (colour, creed, a ban on children and pets, *etc*) and favour the young or mobile in order to reduce the risks of indefinite security of tenure. Nevertheless, without government provision of housing, S_1T families will still go homeless or be forced to share housing in extended family groups.

B Control rents at P_3. A fair rent set at P_3 would represent market equilibrium in the absence of the 'temporary' supply constraint. Moreover, it has the advantage that it offers landlords some inducement to expand supplies in the long run under increasing cost conditions and consumers are not being asked to pay more then the 'normal' long-run cost. On the other hand, an unmet need of S_1Z units would still remain in the short run but falling to zero in the long run.

C Permit rents to rise to P_2. This would mean effective abandonment of fair rents in the short run, given the expectation that in the long run equilibrium would be achieved at P_3. In the short run, landlords would gain considerable surplus but, given freedom of entry, there would apparently be a strong incentive to supply OZ units. No rationing other than ability and willingness to pay is necessary or probable but, even in the long run, there would remain unmet need of ZT.

D Subsidise demand. Under this strategy, the state could abolish rent controls but accept that 'needs' outstrip ability to pay and so subsidise demand either at a flat rate per unit of housing or through an *ad valorem* subsidy. In Figure 8.3, the removal of controls at P_1 and the subsidising of demand by P_2P_4 per unit would, in the short run, increase rents to P_4. In the long run the shift of the demand curve to the right would yield a new equilibrium level which coincided with need of OT units at the rent of P_5. Both DD and D_1D_1 demand curves will 'kink' at OT units for the same reasons given under Figure 8.2, except that in the case of D_1D_1 it is assumed that every family with assistance can pay a rent of P_5. Demand, therefore, is perfectly inelastic up to and including the fair rent level of P_5. In such circumstances, the state would pay the

Figure 8.3: The effects of housing subsidies

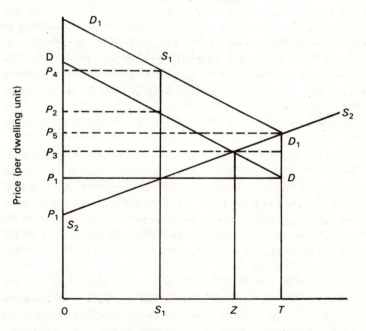

Dwelling units for rent (per time period)

121

difference between the tenant's payment of rent P_1 and the landlord's receipt of P_5. The relative gains to the tenant and the landlord will depend upon the elasticities of supply and demand. In Figure 8.3, the tenant is now paying P_1P_3 less than he would have in the former equilibrium position and the landlord is receiving P_3P_5 more. The greater the elasticity of supply and the lower the elasticity of demand, the greater will be the relative gain to the tenant from any given subsidy, and vice versa. In the case of an *ad valorem* subsidy, the gap between the two demand curves would widen or narrow depending upon whether the subsidy was progressive or regressive.

The relative success of any one of these options depends on the supply and demand conditions, demographic factors, other tenure competition, *etc*. It is not surprising that recent policy has attempted to encourage increased supply outside the strict provisions of rent regulation. Even so, landlords' anticipation of future control of rents could be counterproductive to measures of decontrol.

EXPANSION OF THE VOLUNTARY SECTOR

Recently there have been a number of government initiatives to encourage public and private bodies to invest in rented housing. The types of bodies that could be interested have included development companies, building societies and housing associations. In some cases, such as an extended role for housing associations, the public finance costs could be considerable.

Within the Housing Act 1980, an attempt was made to encourage the provision of rented housing by financial institutions through the introduction of the assured tenancy scheme. Landlords approved by the government have been permitted to build new housing to let at rents and on terms freely negotiated in the market but subject to the Landlord and Tenant Act 1954. The intention has been to encourage housing associations to apply for registration as approved bodies. The scheme had little success until 1986. Recently, more housing associations have expressed interest and the largest recent participant, Abbey Housing Association, provided some 70 lettings out of a total of about 700 in 1986. The government expects the number to grow to 1,500 by 1988. But progress is still slow and the scheme has still had little impact on the rented sector, even though it is outside normal rent regulation. Under the Housing Act 1986 there has been a further extension of the approved landlord scheme. It allows the refurbishment of empty dwellings to approved standards by approved landlords and let on assured tenancies, offering security of tenure, subject to five-year rent reviews, just like most commercial leases. Such an initiative begs the question of whether the government might soon abolish the right of secure tenants to hand on tenancies to their children. This would imply that the pace of renewal of the rented sector would be set by the life expectancy of existing tenants and the supply of approvable landlords. Although these are reforms which may improve access to and supply of rented accommodation, it must be said that the 1986 Housing Act has not gone far enough to arrest the continuing decline of rented housing.

ABOLITION OF RENT AND EVICTION CONTROLS

In addition to the assured tenancy scheme, the Housing Act 1980 also introduces the shorthold tenancy scheme. Under such a tenancy a landlord is able to gain repossession of rented property after a specified time. The minimum period is one year. Originally, it had been a stipulation that a fair rent should be applied to shorthold tenancies. This was later waived, except in London. While tenants are able to appeal to a rent officer it is no longer necessary to register a fair rent at the commencement of a shorthold tenancy. Although these new shorthold tenancies have overcome two major impediments to landlord letting — full security of tenure and potential capital loss — they have not been sufficient to induce a large-scale return of landlords to the housing market. The fact that a landlord no longer faces risks of considerable capital loss is certainly a major attraction. In non-shorthold tenancies, a tenant can apply successfully to a rent officer for a fair rent to be set and full security of tenure.

Todd's (1986) survey of private lettings between 1982 and 1984 yielded a number of interesting results: (a) Among respondents to the survey, only four per cent of tenants against 14 per cent of landlords of recent lettings considered their tenancies to be shorthold; (b) Conversely, 21 per cent of tenants against eight per cent of landlords believed themselves to enjoy full tenure protection within the stated fixed term.

While both shorthold and assured tenancies have provided some little move towards deregulation, their application is still not widespread. They have done little, to date, to reverse the trend in the decline of the rented sector. In London, licences have been used proportionately more than elsewhere, while the use of shorthold tenancies appears to be proportionately greater outside London.

The Housing Act 1980, and its new forms of tenure have had some effect in cushioning a worsening situation but its impact should not be overstated. For example, the rent acts have encouraged the letting of accommodation to the young and mobile. The fact that the Housing Act 1980 has encouraged a further boost to this trend owes less to the intrinsic virtues of the act than to the release of a valve from the pressure arising from pernicious and counterproductive rent acts (Albon and Stafford, 1987).

Under the Housing Act 1986, the government lifted two restrictions on private landlords as part of the government's right-to-rent campaign. Landlords may now impose higher rents, as agreed by a rent officer, immediately instead of having to phase them over two years. Moreover, landlords in Greater London are able to let property on a yearly basis without needing to register a fair rent with a rent officer.

However, the real problem is that all the above proposals, ranging from municipalisation to extension and/or modification of the present acts, fail to overcome the allocation damage of the main prevailing system of rent regulation. The Housing Act 1980, which introduced shorthold tenancies, has had virtually no effect on the decline in this sector and the assured tenancy scheme has been slow. Theoretical analysis, elaborated earlier, makes quite clear that under the present control of rents no increase in the supply of rental housing can be forthcoming.

A repeal of the rent acts and less government intervention in the rented sector is necessary if long-term revisions are to be implemented and a free market allowed to re-emerge after an absence of 70 years. It is clear that the risks associated with present controls continue to induce landlords to cease letting at the earliest opportunity. Those who continue to let may attempt to minimise the risk of financial loss by letting to the more transient groups of tenants — students, holidaymakers, *etc* — and by reducing the standard of maintenance of their property.

Britain now stands out from most countries in retaining a system of rigid rental market controls that have paralysed the private rental housing sector. While other countries have removed their controls, Britain has not. Moves in this direction in the late 1950s were reversed in the mid-1960s. Various schemes have been devised over the years to lighten the load on tenants by either replacing subsidies from landlords with those from the state or by graduating decontrol

124

over a period of time.

An interesting recent proposal for decontrol was suggested by Martin Ricketts (1986). It involves giving sitting tenants clearly-defined property rights to the stream of subsidy from rent control for a period of, say, ten years. Tenants would have the right to remain in their property at a controlled rent for the nominated period *or* sell this right to whomever they chose (*e.g.* another tenant, the landlord or a housing association) for a lump sum. All new tenancies would be decontrolled completely. The scheme is designed to enhance mobility, restore the incentive to invest in rental housing, partially protect the tenant's interest and ultimately result in the complete end of rent and eviction controls.

Similar schemes to aid the transition from a regime of rent control to one of a free market have been proposed in the past. Harrod (1947) suggested that sitting tenants be entitled to the difference between market and controlled rents for a period of ten years whether they stayed in the dwelling or not. The landlord would be taxed the rent difference and this amount paid to the tenant for ten years. As in Ricketts' scheme, new tenancies would be decontrolled completely. The only difference is that all amounts are in flows-per-period rather than in capitalised present values. Paish (1950) proposed a variant of Harrod's plan where some of the tax on landlords would not go to sitting tenants but be deployed elsewhere. He proposed that 25 per cent be used for an allowance to landlords to maintain their properties and some of the rest to fund an income maintenance programme.

Albon and Stafford (1987) argued that the Paish variant of Harrod's scheme is preferable to that of Ricketts in a number of respects. First, it has more chance of being embraced warmly by landlords since they would get something out of it, *i.e.* the means to maintain their properties better. Second, it helps tenants as well since they have the benefits of better maintained living conditions. In fact, the stimulus to better maintenance benefits everyone. Third, it creates some funds that could be used to help any real victims of higher free market rents. Albon and Stafford (1987) thought that new tenants, renting at market rates, could be embittered by the generosity of Ricketts' scheme, which had no provision for them. Instead, they have approached reform by separating the issue of income maintenance from that of housing policy.

While the various types of scheme discussed above are motivated by a desire ultimately to end rent control and to reap some immediate benefits the degree of generosity to tenants could be considerable.

125

It should be borne in mind that sitting tenants have already enjoyed a subsidy (sometimes of a large magnitude for a long period of time) at the expense of their hapless landlords. Any enforced transfer has no clear moral justification, particularly when considered in the light of the circumstances of many of the tenants and landlords concerned.

In Albon and Stafford (1987) a matrix of wealth of tenants and landlords makes this clear (Figure 8.4). It is only in category *B* that rent control produces a redistribution with any moral defensibility. In category *C*, poor landlords subsidise rich tenants, and the other categories, involving poor-to-poor (*A*) and rich-to-rich (*D*) subsidies should not be very encouraging to the advocates of rent control. Knowledge of the attributes of landlords and tenants does not encourage the belief that category *C* is the normal pattern of redistribution. Although the number of tenants really deserving of

Figure 8.4: Redistribution from rent control

Landlords

		Poor	Rich
Tenants	Poor	*A*	*B*
	Rich	*C*	*D*

compensation from landlords may be low, the number that society might wish to compensate — at the expense of taxpayers in general — could be quite large. If decontrol did occur 'overnight' or at one fell swoop, with no change to any of the existing subsidy arrangements, many tenants would be automatically compensated by both the current system of housing benefits (many tenants would become eligible when their rents rose) and by some other elements of the welfare system. In fact, only those tenants who were, in a sense, able to afford the market rent would be ineligible for compensation. Albon and Stafford have argued that this would be an improvement, in that the state would take over the responsibility for subsidies from landlords and subsidies would be far better directed. Aggregate increase in subsidy payments would be less than the transfer from tenants to landlords. Such a transfer of the burden of subsidy from landlords to government is justified. After all, the housing benefit system is supposed to be capable, technically and administratively, of providing support to lower-income tenants. By contrast, those who argue that these costs are too high for government to bear are merely offering a cynical defence of the *status quo*. Government has the machinery and responsibility to provide help to lower-income households at a level and degree it deems appropriate. It is not the duty or responsibility of the landlord.

The new Conservative government indicated in the 1987 Queen's Speech that while the protection of the rent acts will be maintained for existing tenants, new private tenants will be able to negotiate with prospective landlords free of the prevailing system of rent controls. 'Shortholds' and 'assured tenancies' will, argues the government, give landlords the prospect of a recovery of their properties while still protecting tenants against harassment and illegal action. It remains to be seen whether this really does open up the private rental market. Previous such half-way houses introduced by Conservative administrations have failed to do so, in part because landlords could not be certain that a future Labour government would not reverse the legislation. The role of expectations about future controls/decontrols is important. It can be argued that even total decontrol will not promote much investment in housing to rent if the investors think that there is any chance that a future government will impose new controls. Alternatively, even if a present government keeps controls, one might expect investment to start if landlords came to expect them to be removed. In political terms, this seems to argue that only a consensus policy could be relied on to revive investment. Such a consensus seems unlikely to be realised.

A further aspect is the possibility of gradual decontrol. For example, allowing controlled rents to rise by inflation + n per cent each year until they became inoperative. An argument against this is that the mobility effects of controls would be removed only gradually. Also it might be believed that the continued existence of the machinery of control would make a reversal of the policy easier. Against this could be put the argument that many people are genuinely worried that the immediate removal of controls would cause a massive rush of homelessness. Gradual decontrol would give time to demonstrate that this was not happening, and one might even hope for example that, as with the case of council house sales, after a few years the benefits would have become sufficiently obvious that even a Labour government would probably not entirely reverse the decontrol process.

9

Conclusions

This book has been critical of the many long years of government intervention in housing. Yet there was a time when the entire housing market in Britain was virtually free of any significant intervention by the government. Gradually — beginning in 1916 — the role of the state has become more and more pervasive with intervention in the owner-occupancy and council housing sectors. Private rentals have also been subsidised directly by government in circumstances where the landlord cannot be called on more than at present. The tendency has been one of subsidising all sectors rather than none. If tenure-neutrality has ever been an aim it has not been achieved because the rates of subsidy differ markedly within and between tenure types. Regulation has stifled exchange and choice and hindered tenant mobility (Minford, Ashton and Peel, 1987).

COUNCIL HOUSING

As the private rented sector has been gradually killed, the task of providing rental accommodation to lower-income people at below-market rents has been taken over by local authorities. At the end of the 1970s council housing accounted for almost a third of the total housing stock. After several years of council house sales it is still over a quarter. There is a higher proportion of public sector housing in Britain than in other comparable countries where policies have been much more conducive to private rental. In 1987–8 total public sector housing spending on local authority housing is estimated at £2,500m. Moreover council tenants are also substantial beneficiaries from housing benefit, which has risen from £400m in 1981–2 to over £3,200m in 1987–8, at which level it is projected

129

to remain for this decade. Council tenants are therefore not only subsidised by virtue of the nominal rent levied on the property in which they live, but also as beneficiaries of housing benefit according to the relationship between their personal income and this nominal rent. These enormous state subsidies were neither intended nor anticipated in the past. Although council rents increased by 130 per cent between 1979 and 1984 (after years of virtually no increase at all notwithstanding rapid inflation), council tenants continue to spend a far smaller proportion of their income on housing than any other tenure group. For example, in 1983 average unrebated rents paid to local authorities were five per cent lower than housing association tenancies, and 39 per cent lower than regulated furnished tenancies.

Local housing authorities are still allowed to fix their own rents. Many Labour-controlled councils raise cash from ratepayers to keep rents to a minimum. The consequence has been a wide disparity in rents. For example, in 1985, two London boroughs were charging £13 and £24.50 per week respectively for a similar three-bedroomed house. In England and Wales outside Greater London, the unrebated rent for a three-bedroomed council house ranged from £22 to £10 a week in 1985. Rents are therefore not only very low but vary greatly between authorities. Earlier the discrimination that exists in the private rented sector where rents are controlled was noted. It is also clear that in the public sector, rent levels depend less on the type and quality of house than on the political colour of the housing authority. Of course, any increase in rents, either in the public or private sectors, may mean the Exchequer paying yet more than the present £3,200m plus. At present some 65 per cent of all council tenants receive housing benefit and half of those are on supplementary benefit and so have their rent paid in full.

Sales of council houses to tenants able and willing to buy them should continue. Greater flexibility in management and ownership of estates have great advantages. In short, if the public housing sector was smaller it might be better managed. This process of accelerated privatisation as outlined in the Queen's Speech 1987 does not, of course, conflict with arguments that some public sector housing will always be needed to provide for the special needs of those who through age or disabilities are unsuited to managing their own accommodation.

OWNER-OCCUPIED HOUSING

Mortgage interest tax deductibility

In Chapter 6, it was argued that where there is a comprehensive income tax system in operation it is proper to tax the imputed income from owner-occupied housing and allow mortgage interest payments as a deduction from taxable income. This used to be the case until 1963 even though, for many years prior to that, the Inland Revenue had been collecting revenue on a base that was much less than the actual imputed income. In 1963, the then Conservative government made a serious mistake — it removed entirely the need to include imputed income on taxable income without also abolishing mortgage interest deductibility. Income tax relief on qualifying interest on loans for the purchase or improvement of owner-occupied property amounted to approximately £4,750m in 1985–6. The recent proposals of the Labour party and the Alliance to limit this relief to standard rate is misplaced. The proper reform is to retain mortgage interest relief but re-introduce the tax on imputed net income.

Exemption from capital gains taxation

A landlord is subject to taxation on his net income from property letting and to capital gains taxation. The owner-occupier, on the other hand, is not only exempt from imputed income taxation but from capital gains taxation. The tax structure strongly favours owner-occupiers and disfavours landlords in all major respects. This bias is totally unjustified on either equity or efficiency grounds and should be corrected.

PRIVATE RENTED HOUSING

There is no doubt that the demise of the private rental housing market is, in large part, the consequence of government interference not only in this sector directly but also as argued above in regard to owner-occupancy and in the provision of council housing. In the case of owner-occupancy and council housing the source of the subsidies is taxpayers at large while in the case of private renting it is the landlord. This feature is of considerable importance in

131

explaining the dramatic decline in private renting in Britain. Maybe the state has a role in advising tenants and landlords about the pitfalls that might be encountered in formulating contracts. This could be achieved by establishing a tenants' and landlords' advisory service with a tribunal attached. The tribunal could have some role in setting what are effectively market rents when the parties cannot agree, *e.g.* as arbitrator. The state does have a role in providing legal redress where contracts made between parties have not been honoured and also to protect landlords and tenants alike from the unscrupulous. There is little doubt that the law on harassment under the present rent regulation legislation has not been properly enforced. There are few successful legal routes through which tenants can claim compensation from their former landlord when there has been illegal eviction or forcing out by harassment. The state owes a responsibility to give the courts powers to relate compensation at least to the landlord's financial gain as a result of obtaining vacant possession. However, the state has no need to perpetuate and enforce statutory rent-setting machinery and it is time that rent regulation was abolished.

TENURE NEUTRALITY

The government needs to get out of the rented housing market completely by abolishing rent control, reforming mortgage interest tax deductibility and removing council housing subsidies and encouraging privatisation. Any state 'benefits' should be directed at those specifically on low incomes and not according to housing tenure divorced from income. Government interference in the housing market has been a failure and each attempt to 'improve' the situation seems to make things worse. One approach suggested by Albon and Stafford (1987) might be to go beyond the recently revised housing benefit scheme and to aim for true tenure neutrality where all segments of the market are subsidised at the same rate and at the expense of the taxpayer in general. This could involve *inter alia*, the removal of rent control and its replacement by an across-the-board rent subsidy to tenants paid by the Exchequer. The subsidy would have to be equated with that for council housing, involving an adjustment of rents in that sector to conform with the general rate of subsidy. Also it would have to be combined with a properly targetted income-maintenance scheme available to all those meeting the criteria irrespective of type of housing tenure. Depending on the general rate of housing subsidy chosen and the generosity of income maintenance, this scheme may or may not result in an increase in public outlays, including tax expenditures.

Albon and Stafford (1987) have also postulated a more radical approach towards achieving tenure neutrality. This would be to set the across-the-board housing subsidy at precisely zero! How would this impact on the different tenures if the reforms envisaged above were implemented?

Private rents would almost all rise at least in the short run by amounts ranging from very little in some cases to a great deal in others. Virtually all private tenants would pay more with some receiving a double shock where rent subsidies were previously paid. Compensation would occur for those that were eligible for general income maintenance, but many would pay more in an outright sense — perhaps much more. Landlords would definitely gain but tax-payers in general would lose if these increased income maintenance payments exceeded savings on rent subsidies. Council house rents would also rise, again by varying amounts but probably more uniformly. Housing benefits would also be removed, but many would receive compensation under income maintenance and the effect on taxpayers would be ambiguous, but probably negative. A re-introduction of taxation on the net imputed income from owner-occupation (*i.e.* mortgage interest deductibility would be allowed as a housing cost) could raise the housing costs of many owner-occupiers buying their homes and may have a downward effect on property values. Taxpayers would almost certainly gain in net terms as savings in tax expenditures would probably exceed any income maintenance payments to lower income groups in this sector.

Whether society would gain overall from the removal of all housing subsidies depends on assumptions. In previous chapters the costs in economic efficiency associated with rent control have been highlighted. Similar efficiency gains would flow from the abolition of other subsidies which distort production and consumption deci-sions, impede mobility, *etc*. Thousands of millions of pounds worth of extra output could flow from the enhancement of labour mobility alone (see Minford *et al*, 1986 and 1987). Gainers could easily compensate the losers from this set of changes with a social gain left over. Of course, society may not wish to compensate all the losers from the change. The income maintenance scheme would mean that subsidies were directed where they were most needed.

Political repercussions of such radical changes may not be as considerable as individual tinkering with particular parts of the subsidy apparatus. Small attempts in the past to decontrol the rental housing market or reform the extent of mortgage interest deductibility have been met with howls of protest from those affected. Even council house sales which are now an ordinary transaction of authorities were subject initially to tremendous obstruction and opposition in the early 1980s. From the standpoint of any group affected by changes, their degree of acceptance is dependent on both the respective costs and benefits. While there may be some short-term loss of subsidy, it is clear that there would be considerable benefits from major reform and state disengagement in the housing market. From the viewpoint of any recipients, their degree of acceptance of the loss of their subsidy may at worst be enhanced by the knowledge that others are also losing. Threatening only one group can be seen as being horizontally inequitable.

Bibliography

Albon, R.P. (1980) (ed) *Rent control: costs and consequences*, Centre for Independent Studies, Sydney
—— (1981) *An economic evaluation of rent control and other rental housing market policies in Australia*, doctoral thesis, Department of Economics, Australian National University, Canberra
—— and Stafford, D.C. (1987) *Rent control*, Croom Helm, London
Arden, A. (1980) *The Housing Act 1980*, Sweet and Maxwell, London
Audit Commission (1986) *Managing the crisis in council housing*, Audit Commission for Local Authorities in England and Wales, Her Majesty's Stationery Office (HMSO)
Aughton, H. (1981) *Housing finance: a basic guide*, Shelter, London
Bank of New South Wales (1953) 'Implications of rent control', *Bank of New South Wales Review*, 12 February, 10–13, reprinted in R.P. Albon (ed), Centre for Independent Studies, Sydney, ch. 7 (ii) (1980)
Begg, D., Fischer, S. and Dornbusch, R. (1984) *Economics*, British edition, McGraw-Hill, London
Boleat, M. (1986) *The building society industry*, Allen and Unwin, London
Browning, E.K. and Browning, J.M. (1983) *Microeconomic theory and applications*, Little, Brown and Company, Boston
Call, S.T. and Holahan, W.L. (1977) *Microeconomics*, 2nd edn, Wadsworth, Belmont, California
Cheung, S.N.S., (1980) 'Rush or delay: the effects of rent control on urban renewal in Hong Kong'; in R.P. Albon (ed) *Rent control*, ch. 2 (1980)
Clyne, P. (1970) *Practical guide to tenancy law*, Rydge Publications, Sydney
Commission of Inquiry into Poverty (1975) (R.F. Henderson, chairman) *First main report*, vols 1 and 2, Commonwealth of Australia, Canberra
Cooper, M.H. and Stafford, D.C. (1975) 'A note on the economic implications of fair rents', *Social and Economic Administration*, 9, 1, spring, 26–29
—— and —— (1979) 'The economic implications of fair rents'; in R.A.B. Leaper (ed), *Health, wealth and happiness*, Blackwells, Oxford
—— and —— (1980) 'Rent control: the United Kingdom experience'; in R.P. Albon (ed) *Rent control*, ch. 4 (1980)
Cullingworth, J.B. (1965) *English housing trends*, occasional papers on social administration no. 13, Bell and Sons, London
—— (1969) *Housing and labour mobility: a preliminary report*, OECD, Paris
—— (1979) *Essays on housing policy: the British scene*, Allen and Unwin, London
de Jouvenal, B. (1948) 'No vacancies'; reprinted in *Verdict on rent control*, Institute of Economic Affairs, Readings No. 7, London (1972)
De Leeuw, F. (1971) 'The demand for housing: a review of cross-section evidence', *The Review of Economics and Statistics*, vol. 53, No. 1
—— and Ekanem, N.F. (1971), 'The supply of rental housing',

American Economic Review, 61, 5, December, 806–17

Department of the Environment (1970) *National dwelling and housing survey*, HMSO

────── (1977) *The review of the rent acts*, consultation paper, HMSO

Frankena, M.W. (1975) 'Alternative models of rent control', *Urban Studies*, 12, 3, October, 303–8

Fraser Institute (1975) *Rent control: a popular paradox*, Vancouver, Canada

Friedman, M. (1962) *Capitalism and freedom*, University of Chicago Press, Chicago

────── and Stigler, G. (1946) *Roofs or ceilings?* Foundation for Economic Education Inc, Irvington-on-Hudson, New York; reprinted in *Verdict on rent control*, Institute of Economic Affairs, Readings no. 7, London, (1972); reprinted in R.P. Albon (ed) *Rent control*, ch. 1, (1980)

Gray, H. (1968) *The cost of council housing*, research monograph no. 18, Institute of Economic Affairs, London

Greve, J. (1965) *Private landlords in England*, occasional paper on social administration no. 16, Bell and Sons, London

Harloe, M. (1985) *Private rented housing in the United States and Europe*, Croom Helm, London

Harrod, R.F. (1947) *Are these hardships necessary?*, Rupert Hart-Davis, London

Hayek, F.A. (1929) 'The repercussions of rent restrictions'; reprinted in *Verdict on rent control*, Institute of Economic Affairs, Readings no. 7, London (1972)

Hemming, M.F.W. and Duffy, H. (1964) 'The price of accommodation', *National Institute Economic Review*, no. 29

Henney, A. (1975) 'The implications of the Rent Act 1974', *Housing Review*, March–April

Hepworth, N.P. (1975) 'Local government and housing finance', *Housing Finance*, Institute for Fiscal Studies, no. 12

Hepworth, N.P. (1976) *The finance of local government*, Allen and Unwin, London

Hirshleifer, J. (1984) *Price theory and applications*, 3rd edn, Prentice Hall, Englewood Cliffs, New Jersey

Hughes, G.A. (1980) 'Housing and the tax system'; in Hughes, G.A. and Heal, G.M. (eds) *Public policy and the tax system*, Allen and Unwin, London

────── (1987) 'Rates reform and the housing market'; in Bailey, S.J. and Paddison, R. (eds) *The reform of local government finance in Britain*, Croom Helm, London

Inquiry into British Housing (1985) (HRH The Duke of Edinburgh, chairman), National Federation of Housing Associations, London

Institute of Economic Affairs (1972) *Verdict on rent control*, London

Institute of Public Affairs (1954) 'Rent control', *The IPA Review*, 8, 3, 73–9, reprinted in R.P. Albon (ed), *Rent control*, ch. 7 (i) (1980)

King M.A. and Atkinson, A.B. (1980) 'Housing policy, taxation and reform', *Midland Bank Review*

Lindbeck, A. (1967) 'Rent control as an instrument of housing policy'; in A.A. Nevitt (ed), *The economic problems of housing*, Macmillan, London

McCloskey, D. (1985) *The applied theory of price*, 2nd edn, Macmillan, New York

Maclennan, D. (1978) 'The 1974 rent act — some short-run supply effect', *The Economic Journal*, vol. 88

—— (1980) *Housing economics*, Longman, London

McKenzie, R.B. and Tullock, G. (1978) *Modern political economy*, McGraw-Hill, New York

Malpass, P. (1986) *The housing crisis*, Croom Helm, London

Matthews, R. (1983) *Restrictive practices: waiting list restrictions and housing needs*, Shelter, London

Minford, A.P.L. (1983) 'Labour market equilibrium in an open economy', *Oxford Economic Papers*, November supplement on unemployment, vol. 35 (4); reprinted in *The causes of unemployment* (C.A. Greenhalgh, P.R.G. Layard and A.J. Oswald (eds)), Oxford University Press (1983), pp.207–44

—— Davies, D.H., Peel, M.J. and Sprague, A. (1983) *Unemployment — cause and cure*, 1st edn, Martin Robertson, Oxford

—— Ashton, P., Davies, D.H., Peel, M.J. and Sprague, A. (1985) *Unemployment — cause and cure*, 2nd edn, Blackwells, Oxford

—— Ashton, P. and Peel, M. (1986) *The effects of housing distortions on unemployment*, mimeo, Department of Economic and Business Studies, University of Liverpool

—— Peel, M. and Ashton, P. (1987) *The housing morass*, Hobart paper no. 25, Institute of Economic Affairs, London

Murie, A. (1976) *The sale of council houses: a study in social policy*, occasional paper no. 35, Centre for Urban and Regional Studies, University of Birmingham

—— Niner, P. and Watson, W. (1976) *Housing policy and the housing system*, Allen and Unwin, London

National Consumer Council (1984) *Moving home: why is it difficult for council tenants?*, National Consumer Council, London

Odling-Smee, J.C. (1975) *The impact of the fiscal system on different tenure sectors*, publication no. 12, Institute for Fiscal Studies, London

Olsen, E.O. (1969a) 'A competitive theory of the housing market', *American Economic Review*, 59, September, 612–21

—— (1969b) 'The effects of a simple rent control scheme in a competitive housing market', *Rand Corporation*, New York 4257

—— (1971) 'Subsidized housing in a competitive market: reply', *American Economic Review*, 61, 1, March, 220–4

—— (1972) 'An econometric analysis of rent control', *Journal of Political Economy*, 80, 6, November–December, 1081–100

Paish, F.W. (1952) 'The economics of rent restriction', *Lloyds Bank Review*, April

Paley, B. (1978) *Attitudes to letting in 1976*, Office of Population Censuses and Surveys, Social Survey Division, JS1091, HMSO

Parish, R.M. (1980) 'Economic effects of Victoria's residential tenancies bill'; in R.P. Albon (ed), *Rent control*, ch. 12 (1980)

Parliament of New South Wales (1961) *Report of the Royal Commission of Inquiry on the Landlord and Tenant (Amendment) Act, 1948, as amended*, New South Wales Government Printer, Sydney

Pennance, F.G. (1969) *Housing market analysis and policy*, Hobart paper no. 48, Institute of Economic Affairs, London

—— (1972) *Verdict on rent control*, Institute of Economic Affairs, Readings no. 7, London

—— (1975) 'Recent British experience: a postscript from 1975'; in *Rent control: a popular paradox*, Fraser Institute, Vancouver

—— and Gray, H. (1968) *Choice in housing*, Institute of Economic Affairs, London

Real Estate and Stock Institute of Australia (1975) *Rent controls: study of the effects of rent controls in Canberra*, Canberra

Reynolds, L. (1977) *Some effects of the 1974 Rent Act in London*, Middlesex Polytechnic

Ricketts, M. (1986) *Lets into leases*, Centre for Policy Studies, London

Rydenfelt, S. (1980) 'The rise, fall and revival of Swedish rent control'; in R.P. Albon (ed), *Rent control*, ch. 3 (1980)

Samuelson, P.A. (1980) *Economics*, eleventh edn, McGraw-Hill Kogakusha, Tokyo

Shreiber, C. and Tabriztchi, S. (1976) 'Rent control in New York City: a proposal to improve resource allocation', *Urban Affairs Quarterly*, 1, 4, June, 511–22

Small Landlords Association (1984) *Review of the rent acts*, London

Spicker, P. (1983) *The allocation of council housing*, Shelter, London

Stafford, D.C. (1976) 'The final economic demise of the private landlord', *Social and Economic Administration*, vol. 10, no. 1, spring

—— (1976) 'Government and the housing situation', *National Westminster Bank Review*, November

—— (1978) *The economics of housing policy*, Croom Helm, London

Tingle, R. (1986) *Housing and mobility in Scotland*, Aims of Industry, London

Todd, J.E. (1986) *Recent private lettings 1982–4*, HMSO

—— Bone, M. and Noble, I. (1982) *The privately rented sector in 1978*, Office of Population Censuses and Surveys, Social Survey Division, HMSO

Webster, R.H. (1980) 'Wheeling and dealing under rent control'; in R.P. Albon (ed), *Rent control*, ch. 10 (1980)

Whitehead, C.M.E., Harloe, M. and Bovaird, A. (1985) 'Prospects and strategies for housing in the private rented sector', *Journal of Social Policy*, vol. 14

Wicks, M. (1973) *Rented housing and social ownership*, Fabian Tract, no. 421, Fabian Society, London

HMSO and other official data sources

CIPFA Statistical Information Service, Housing Revenue Account Statistics (various years); Housing Rent Statistics (various years)

Family Expenditure Survey (various years), Department of Employment, HMSO

General Household Survey (various years), Office of Population Censuses and Surveys, Social Survey Division, HMSO

House of Commons 40 (1982): *The private rented housing sector*, vols 1 to 3, House of Commons Environment Committee, HMSO

House of Commons 54 (1982): *Memorandum from the Department of the Environment*, House of Commons Environment Committee, HMSO

House of Commons 201 (1983): *A Report on the Memorandum from the Department of the Environment*, House of Commons Environment Committee, HMSO

HMSO (1931): *Committee on Rent Restrictions Acts*, Cmd. 3911

────── (1965): *Report of the Committee on Housing in Greater London*, (Milner Holland Report), Cmnd. 2605

────── (1971): *Fair deal for housing*, Cmnd. 4728

────── (1971): *Report of the Committee on the Rent Acts* (Francis Report), Cmnd. 4609

────── (1977): *Housing policy: technical volumes 1 to 3* (accompanying Cmnd. 6851), Department of the Environment

────── (1983): *Rates: proposals for rate limitation and reform of the rating system*, Cmnd. 9008

────── (1985): *Reform of social security*, Cmnd. 9691

────── (1986): *Paying for local government*, Cmnd. 9714

Housing and Construction Statistics: Great Britain (various years), Department of the Environment; Scottish Development Department; Welsh Office, HMSO

Office of Population Censuses and Surveys (OPCS) (1982):*Labour Force Survey 1981*, HMSO

────── (1983a): *Census 1981: national migration, Great Britain*, Part 1, HMSO

────── (1983b): *Census 1981: housing and households, England and Wales*, HMSO

────── (1983c): *Recently moving households — a follow up to the 1978 national dwelling and housing survey*, HMSO

Index

Page references in *italics* refer to tables or figures.

age of privately rented housing
68
allocation of housing,
administrative and market 12
allocation rules for local
authority housing 56
annuity mortgages 42
assured tenancy scheme 123

banks, mortgages 40, *41*
baths, fixed, houses without 23,
22
benefit and council rents 50
black markets in rented housing
80
building societies 28, 39
deposits and shares in 40
Building Societies Association 43

capital consumption in housing
17, *18*
capital expenditure on housing
by local authorities 47
capital gains tax exemption of
owner-occupied houses 98
capital receipts of local
authorities, control on use 33
central heating 23, *22*
changes in the demand and
supply of housing 11
changes in tenure 26, *26, 27*
condition of housing stock 21,
22
conditions of resale in council
house sales 110
consequences of rent control 74
controls on local authority
housind expenditure 32
cooperatives, tenants' 105
cost of mortgage interest tax
relief 88, *89*
council house sales 20, 103, *104*
conditions on resale 110

discounts 108
council rents 111
and benefits 50
and subsidies *50*
definition 51

decontrol, gradual 127
demand for housing 7
changes in 11
environmental factors 9
locational factors 8
Department of the Environment
31
deposits in building societies 40
difficult to let dwellings 58
discounts on council house sales
108
discrimination in letting 14
durability of housing 6

endowment mortgages 43
English House Condition Survey,
1981 9
environmental factors in demand
for housing 9
equity sharing schemes 103
Estate Action 26
excess demand for housing 10
for local authority housing 60
Exchequer subsidies 52, *53*

fair rents 64, 118, *119*
filtering down 9
finance for housing 28
first-time buyers 39

gradual decontrol of rents 127

home improvements 19, 23
homeless households, lets to *57*,
58
Homeless Persons Act 58
household size, changes in 20

Housing Act 1980 110
Housing Act 1984 103
Housing Act 1985 33
housing associations 112
housing benefit 29, 32
Housing Corporation 113
housing investment programmes
   of local authorities 31
housing policy, aims 3
housing revenue accounts of
   local authorities 30, 49, *49*
   subsidies from rate funds to
   50
housing services and stocks 7
housing shortage 9, 20
housing standards by tenure *24*
housing subsidies 121, *121*

illegal practices by landlords 81
imputed income from housing 5
   taxation of *92*, 94
income tax, schedule A 29, 93
inheritance and owner-occupation
   39
insurance companies, mortgages
   39, *41*
insurance for repairs 44
investment in housing 17, *18*
   net 17, *18*
   private sector *18*, 19
   public sector 19, *19*

key money 83

landlords
   illegal practices by 81
   tax treatment of 78
leakage of funds from housing
   market 91, *91*
lets to homeless households *57*,
   58
loans for home improvement 23
local authorities
   capital expenditure on
      housing 47
   capital receipts, controls on
      use of 33
   housing expenditure, controls
      on 32
   housing investment

programmes 31
housing revenue accounts 30,
   49, *49*
loans, interest pooling 48
rents 53
   increases 49
   as source of finance 47
locational factors in demand for
   housing 9

maintenance and repairs
   effects of rent control 14
   responsibility for 44
market allocation of housing 12
mobility
   and local authorities' housing
      policies 61
   and preference for renting 44
   rent controls and 82
mortgage interest
   relief at source (MIRAS) 91
   relief, costs of 88, *89*
      £30,000 limit 90
   tax deductibility 28, 92, *92*
mortgages 5
   1971–84 *41*
   annuity 42
   banks 40, *41*
   endowment 43
   insurance companies 39, *41*
   and valuation of houses 43
municipalisation of rented
   housing 116

net investment in housing 17, *18*
new lettings of local authority
   housing 56, *57*
number of dwellings and
   households *18*, 20

overcrowding 23
owner-occupation 3, 37
   drawbacks 44
   in Belgium, Canada, Japan
      and USA 38
   inheritance and 39
   preference for, at different
      ages 37
   and public policy 4
   rise in 26

and shortage of rented
housing 38

points schemes 58
private rented housing
age of 68
decline 27
share in housing stock 63, 67
public policy and owner-
occupation 4
public sector housing 5
aims 46
finance 29, *30*
rise in 27

quality of local authority housing
61

renovation by local authorities 26
Rent Acts 1957, 1964 and 1968
64
Rent Act 1974 65
Rent Acts, review of 1977 65
rent controls 13
consequences 74
and housing quality 77
and maintenance 14
and mobility 82
and supply of rented housing
14
rent increases by local authorities
49
pooling, local authorities 51
rebates *50*
Rent Restriction Acts 63
rented housing 4
black markets in 80
preference for and mobility
44
rent control and supply 14
shortage and owner-
occupation 38
Rents
by tenure 53, *54*
local authority 53
as source of local authority
income 47

repairs and improvements 23
insurance for 44
responsibility for 44
Report of the Enquiry into
British Housing 23
residential qualifications for local
authority housing 58
Review of the Rent Acts 1977
65
right to buy 103

schedule A income tax 29, 93
security of tenure 13
of council tenants 60
shares in building societies 40
shortage, housing 20
shorthold tenancy scheme 123
sitting tenants 13
subsidies on council houses cost
50
definitions 51
on housing 121, *121*
from rate funds to housing
revenue accounts *49*, 50
supply of housing, changes 11

tax deductibility of mortgage
interest 5, 28
tax treatment of landlords 78
taxation of imputed incomes
from housing 5, *92*, 94
tenant cooperatives 105
tenants, unsatisfactory 4
tenure, changes in *26*, 27
security of 13
Thamesmead 106

unfit houses 25
unsatisfactory tenants 4
Urban Housing Renewal Unit
26, 105

valuation of houses and
mortgages 43

waiting lists 58